D1179354

EYEWITNESS VISUAL DICTIONARIES

THE VISUAL DICTIONARY *of* PHYSICS

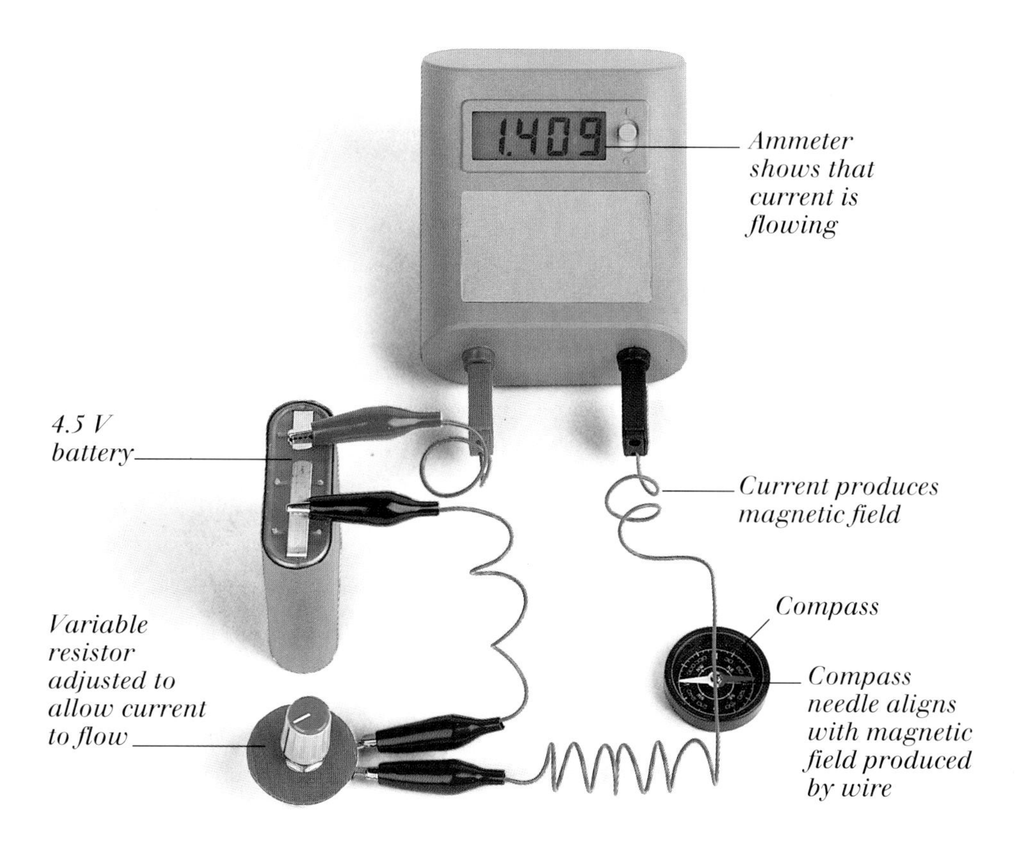

ELECTROMAGNETISM AFFECTING A COMPASS NEEDLE

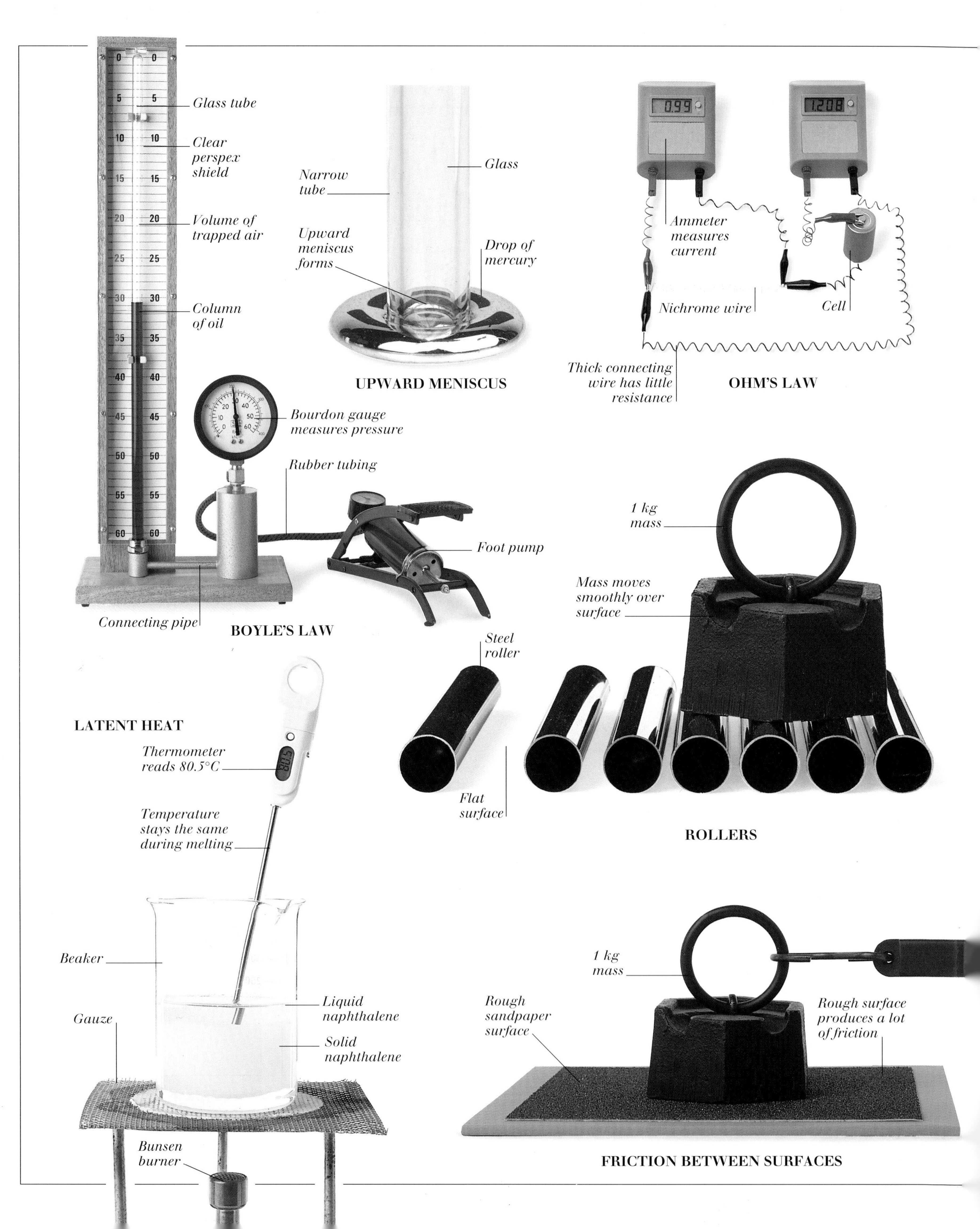
Glass tube
Clear perspex shield
Volume of trapped air
Column of oil
Bourdon gauge measures pressure
Rubber tubing
Foot pump
Connecting pipe
BOYLE'S LAW
Narrow tube
Glass
Upward meniscus forms
Drop of mercury
UPWARD MENISCUS
Ammeter measures current
Nichrome wire
Cell
Thick connecting wire has little resistance
OHM'S LAW
1 kg mass
Mass moves smoothly over surface
Steel roller
Flat surface
ROLLERS
LATENT HEAT
Thermometer reads 80.5°C
Temperature stays the same during melting
Beaker
Gauze
Liquid naphthalene
Solid naphthalene
Bunsen burner
1 kg mass
Rough sandpaper surface
Rough surface produces a lot of friction
FRICTION BETWEEN SURFACES

EYEWITNESS VISUAL DICTIONARIES

THE VISUAL DICTIONARY *of* PHYSICS

written by
Jack Challoner

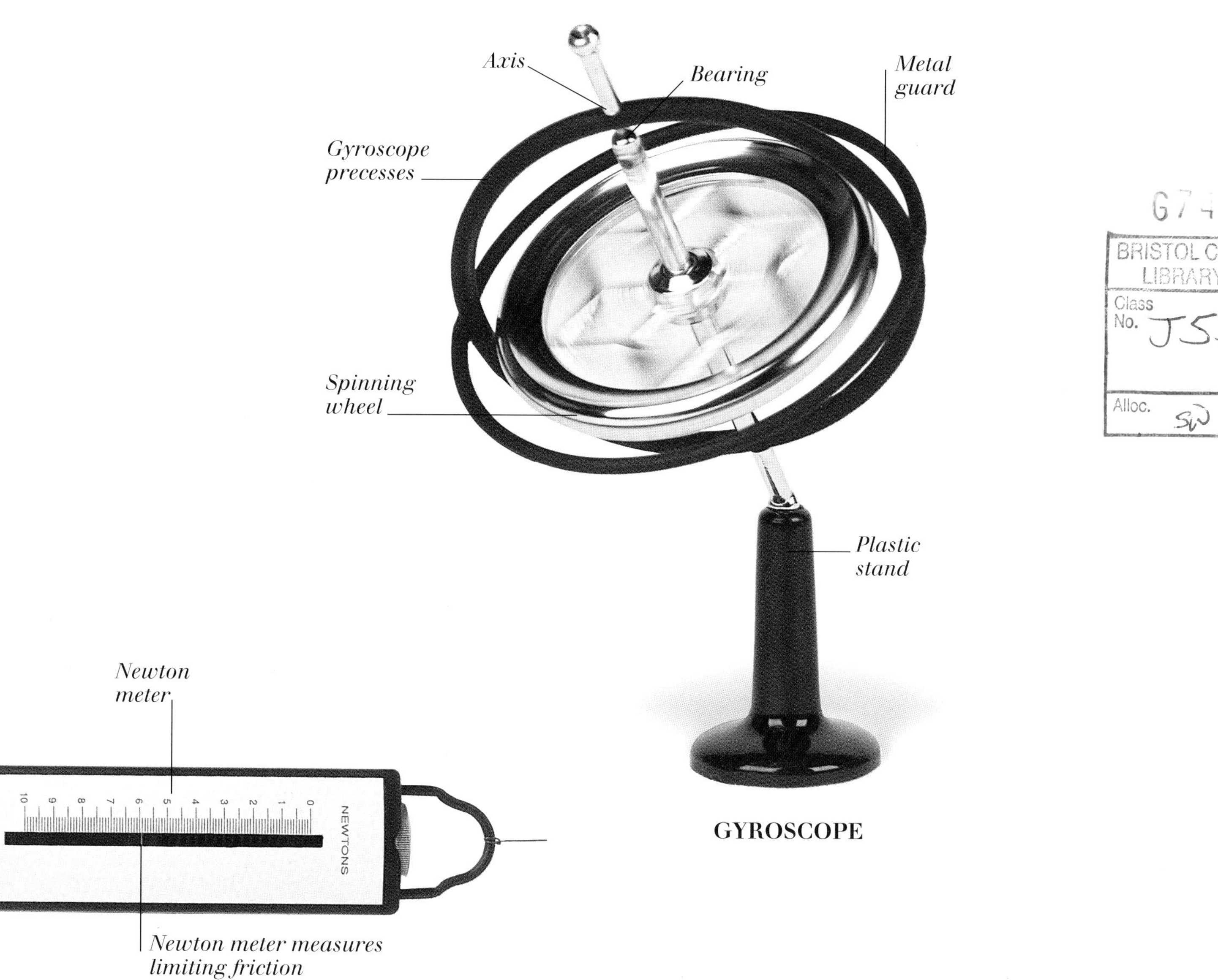

GYROSCOPE

DK
DORLING KINDERSLEY
LONDON • NEW YORK • STUTTGART • MOSCOW

A DORLING KINDERSLEY BOOK

Art Editor Simon Murrell
Project Editor Peter Jones
Editor Des Reid

Deputy Art Director Tina Vaughan
Managing Editor Sean Moore
Senior Art Editor Tracy Hambleton-Miles

Photography Andy Crawford
Illustrations Chris Lyon, Janos Marffy
Picture Research Anna Lord
Production Meryl Silbert

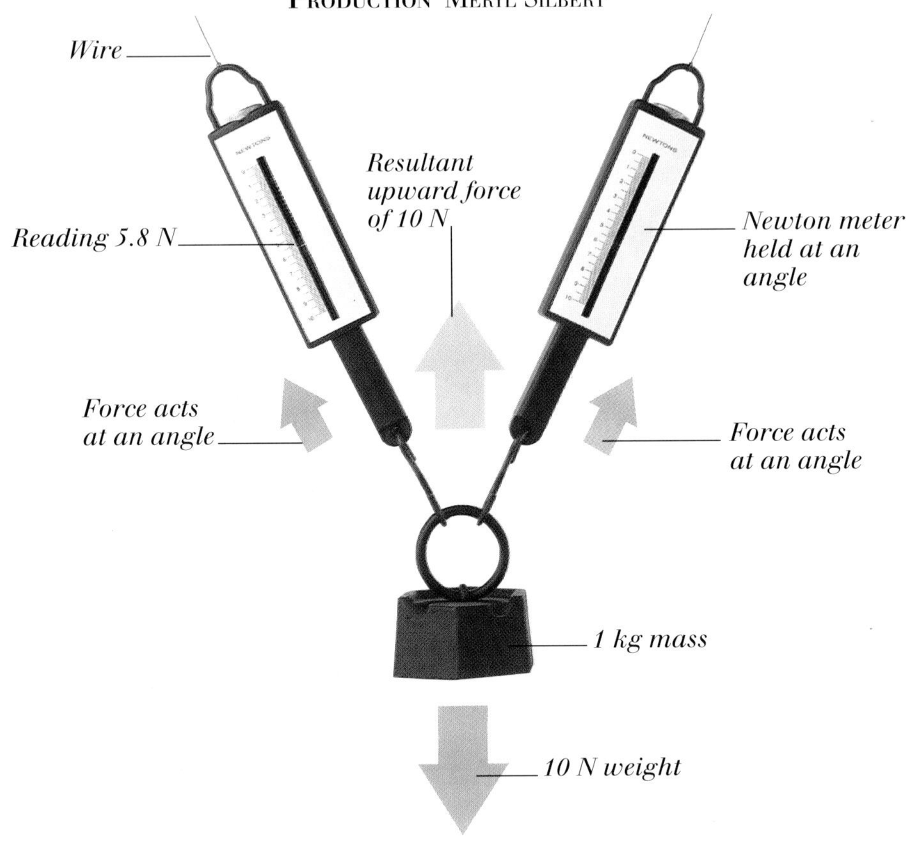

RESULTANT FORCE

First Published in Great Britain in 1995
by Dorling Kindersley Limited,
9 Henrietta Street, London WC2E 8PS

A CIP catalogue record for this book is available from the British Library

ISBN 0 7513 1061 1

Reproduced by Colourscan, Singapore
Printed and bound by Arnoldo Mondadori, Verona, Italy

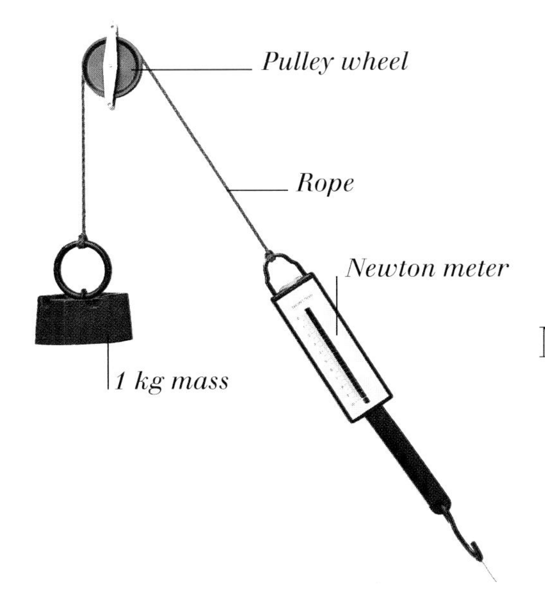

SIMPLE PULLEY

Contents

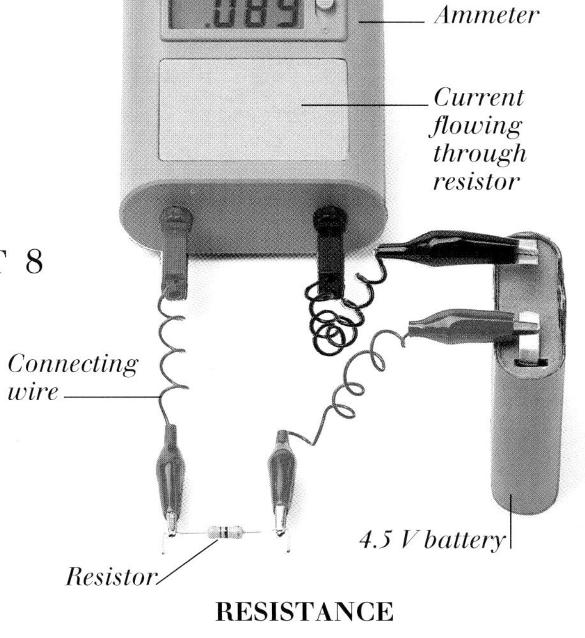

RESISTANCE

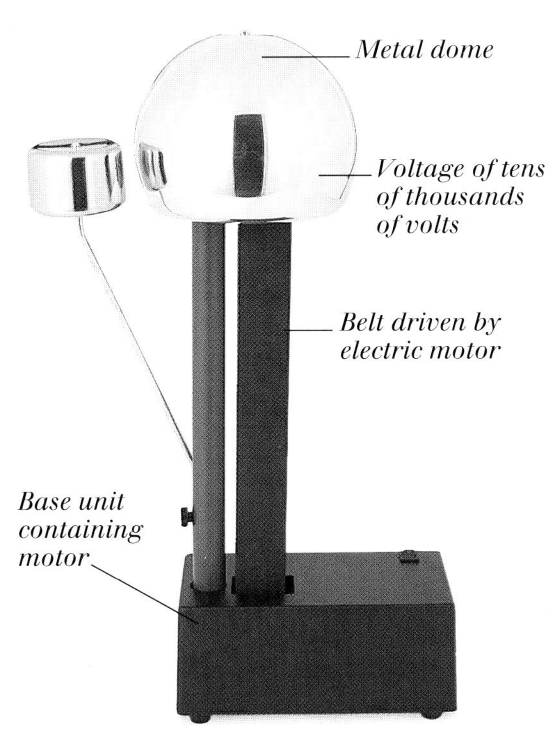

VAN DE GRAAFF GENERATOR

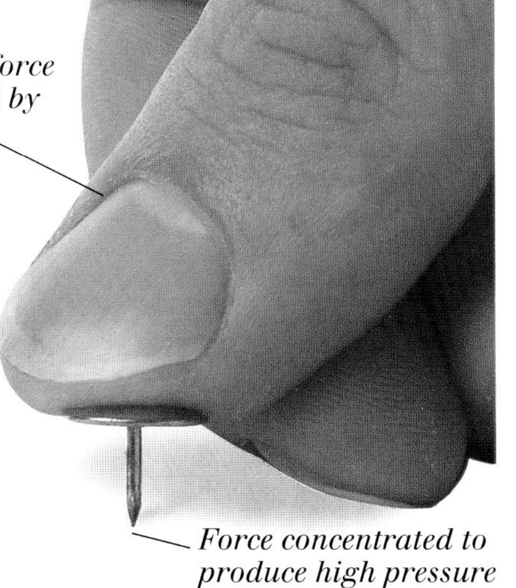

DRAWING PIN

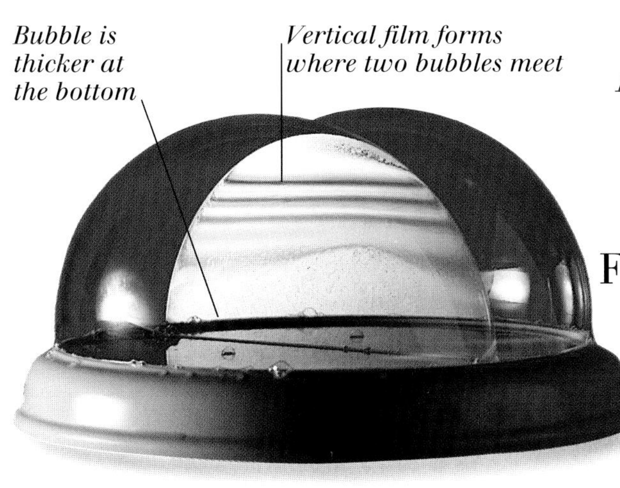

SOAP BUBBLE

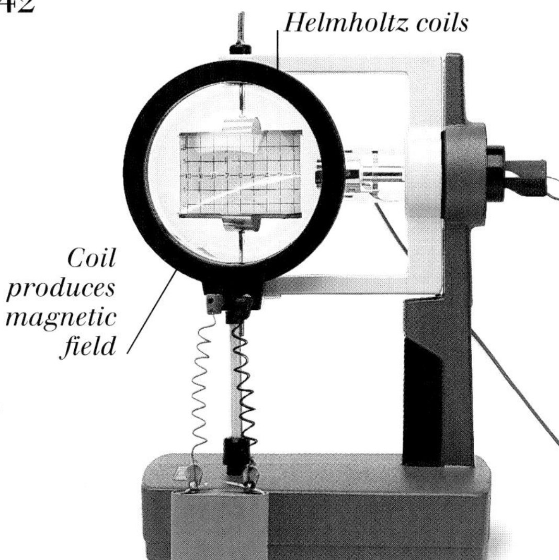

CATHODE RAY TUBE

Matter and energy

PHYSICS IS THE STUDY OF MATTER AND ENERGY. Matter is anything that occupies space. All matter consists of countless tiny particles, called **atoms** (see pp. 48-49) and **molecules**. These particles are in constant motion, a fact that explains a phenomenon known as **Brownian motion**. The existence of these particles also explains **evaporation** and the formation of **crystals** (see pp. 24-25). Energy is not matter, but it affects the behaviour of matter. Everything that happens requires energy, and energy comes in many forms, such as heat, light, electrical, and potential energy. The standard unit for measuring energy is the joule (J). Each form of energy can change into other forms. For example, electrical energy used to make an electric motor turn becomes **kinetic energy** and heat energy (see pp. 22-23). The total amount of energy never changes; it can only be transferred from one form to another, not created or destroyed. This is known as the **Principle of the Conservation of Energy**, and can be illustrated using a **Sankey Diagram** (see opposite).

PARTICLES IN MOTION

BROWNIAN MOTION

When observed through a microscope, smoke particles are seen to move about randomly. This motion is caused by the air molecules around the smoke particles.

SMOKE CELL

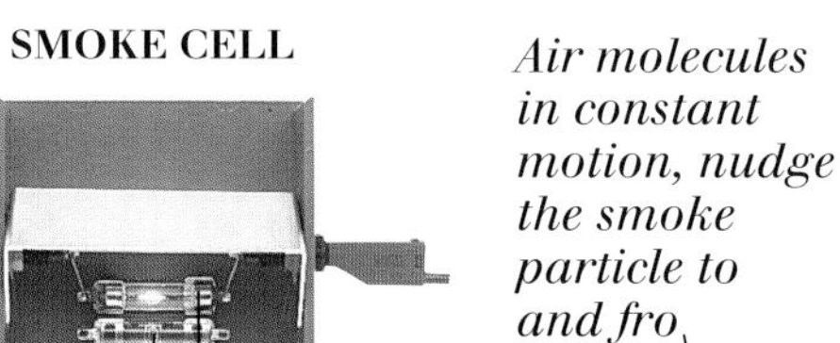

Eyepiece

Objective lens

Smoke cell in place

Wire to battery

MICROSCOPE

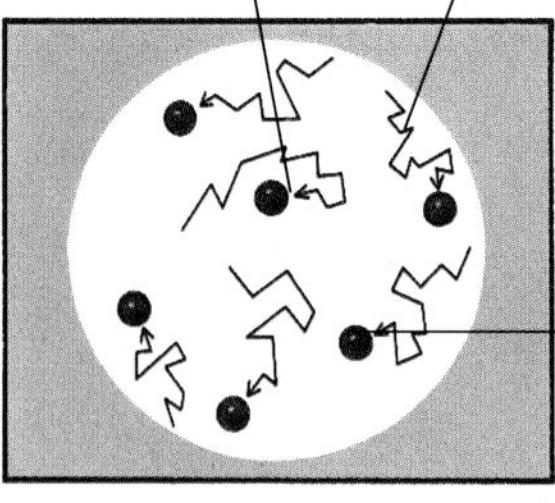

UNDER THE MICROSCOPE

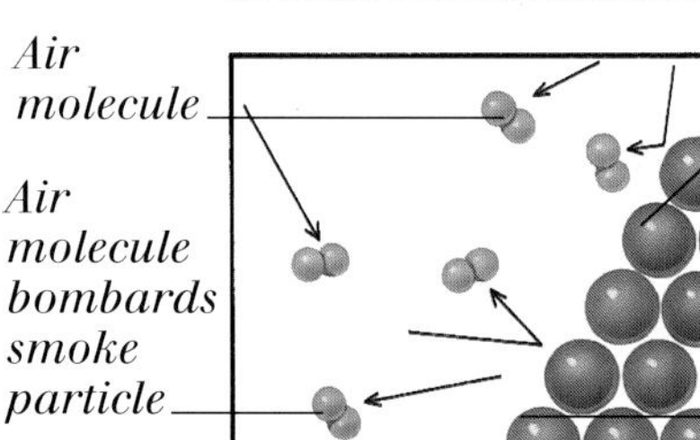

BOMBARDMENT OF SMOKE PARTICLE

MATTER AS PARTICLES

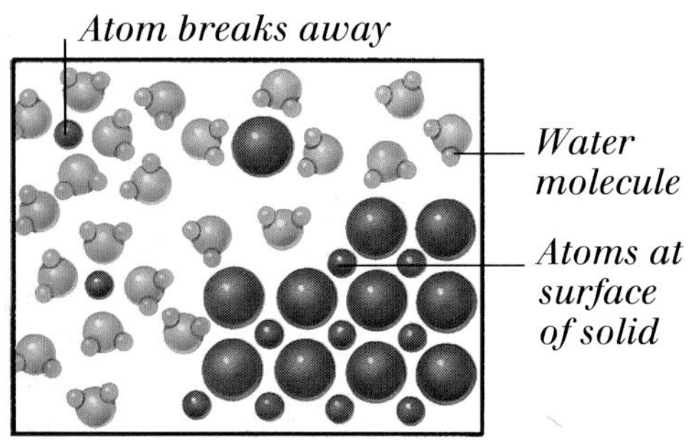

DISSOLVING

The particles of a solid are held together in a rigid structure. When a solid dissolves into a liquid, its particles break away from this structure and mix evenly in the liquid, forming a **solution**.

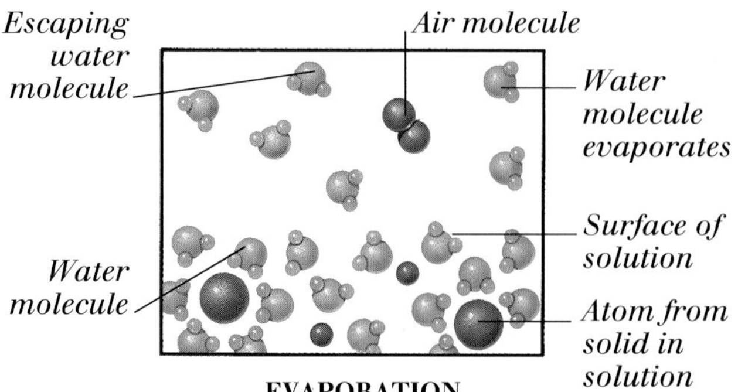

EVAPORATION

When they are heated, most liquids **evaporate**. This means that the atoms or molecules of which they are made break free from the body of the liquid to become gas particles.

Solid particle adds on to structure

Regular crystal structure

Surface of solid

CRYSTALLIZATION

When all of the liquid in a solution has evaporated, the solid is left behind. The particles of the solid normally arrange in a regular structure, called a crystal.

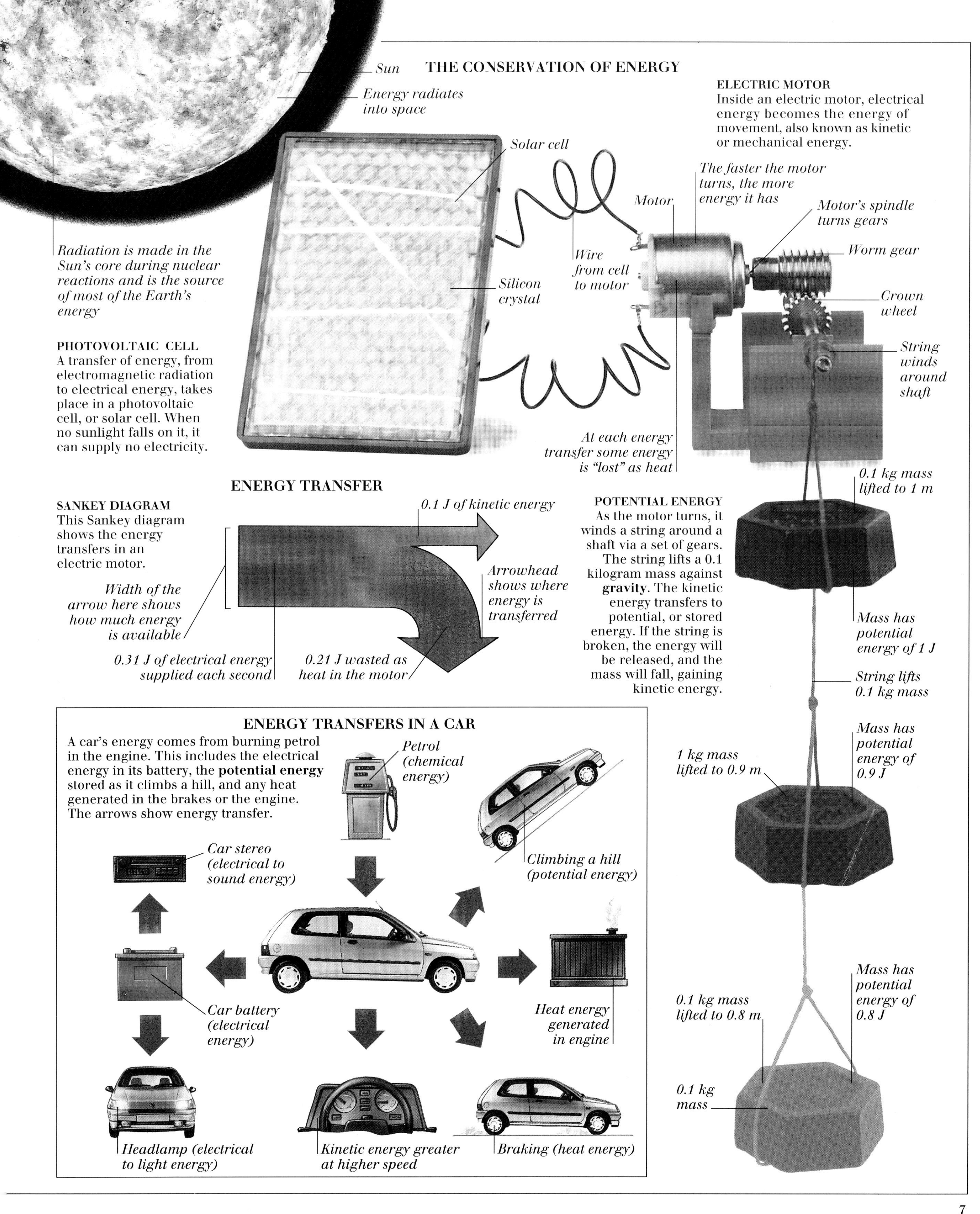

THE CONSERVATION OF ENERGY

ELECTRIC MOTOR
Inside an electric motor, electrical energy becomes the energy of movement, also known as kinetic or mechanical energy.

PHOTOVOLTAIC CELL
A transfer of energy, from electromagnetic radiation to electrical energy, takes place in a photovoltaic cell, or solar cell. When no sunlight falls on it, it can supply no electricity.

ENERGY TRANSFER

SANKEY DIAGRAM
This Sankey diagram shows the energy transfers in an electric motor.

POTENTIAL ENERGY
As the motor turns, it winds a string around a shaft via a set of gears. The string lifts a 0.1 kilogram mass against **gravity**. The kinetic energy transfers to potential, or stored energy. If the string is broken, the energy will be released, and the mass will fall, gaining kinetic energy.

ENERGY TRANSFERS IN A CAR

A car's energy comes from burning petrol in the engine. This includes the electrical energy in its battery, the **potential energy** stored as it climbs a hill, and any heat generated in the brakes or the engine. The arrows show energy transfer.

Measurement and experiment

THE SCIENCE OF PHYSICS IS BASED on the formulation and testing of theories. Experiments are designed to test theories and involve making measurements – of **mass**, length, time, or other quantities. In order to compare the results of various experiments, it is important that there are agreed standard units. The kilogram (kg), the metre (m), and the second (s) are the fundamental units of a system called **SI units** (Système International). Physicists use a large number of instruments for making measurements. Some, like the Vernier callipers, travelling microscopes, and thermometers are common to many laboratories, while others will be made for a particular experiment. The results of measurements are interpreted in many ways, but most often as graphs. Graphs provide a way of illustrating the relationship between two measurements involved in an experiment. For example, in an experiment to investigate falling objects, a graph can show the relationship between the duration and the height of the fall.

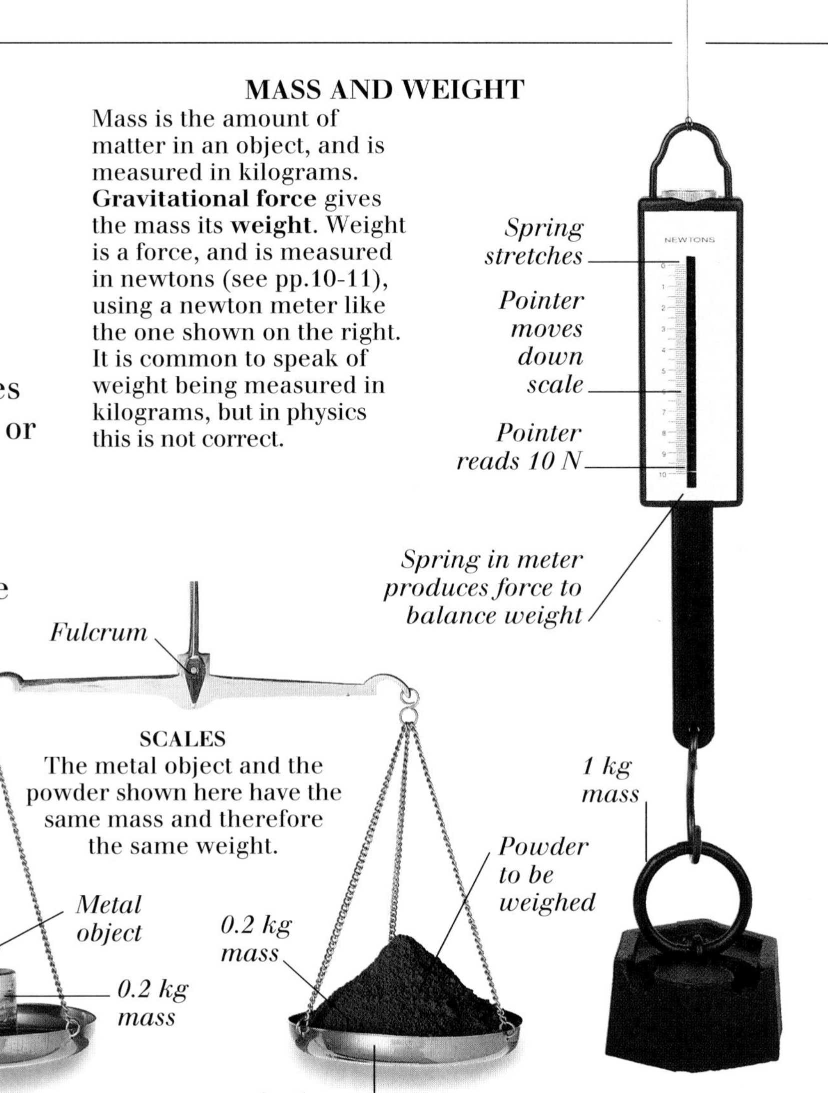

MASS AND WEIGHT

Mass is the amount of matter in an object, and is measured in kilograms. **Gravitational force** gives the mass its **weight**. Weight is a force, and is measured in newtons (see pp.10-11), using a newton meter like the one shown on the right. It is common to speak of weight being measured in kilograms, but in physics this is not correct.

SCALES

The metal object and the powder shown here have the same mass and therefore the same weight.

NEWTON METER AND KILOGRAM MASS

MEASURING DISTANCE

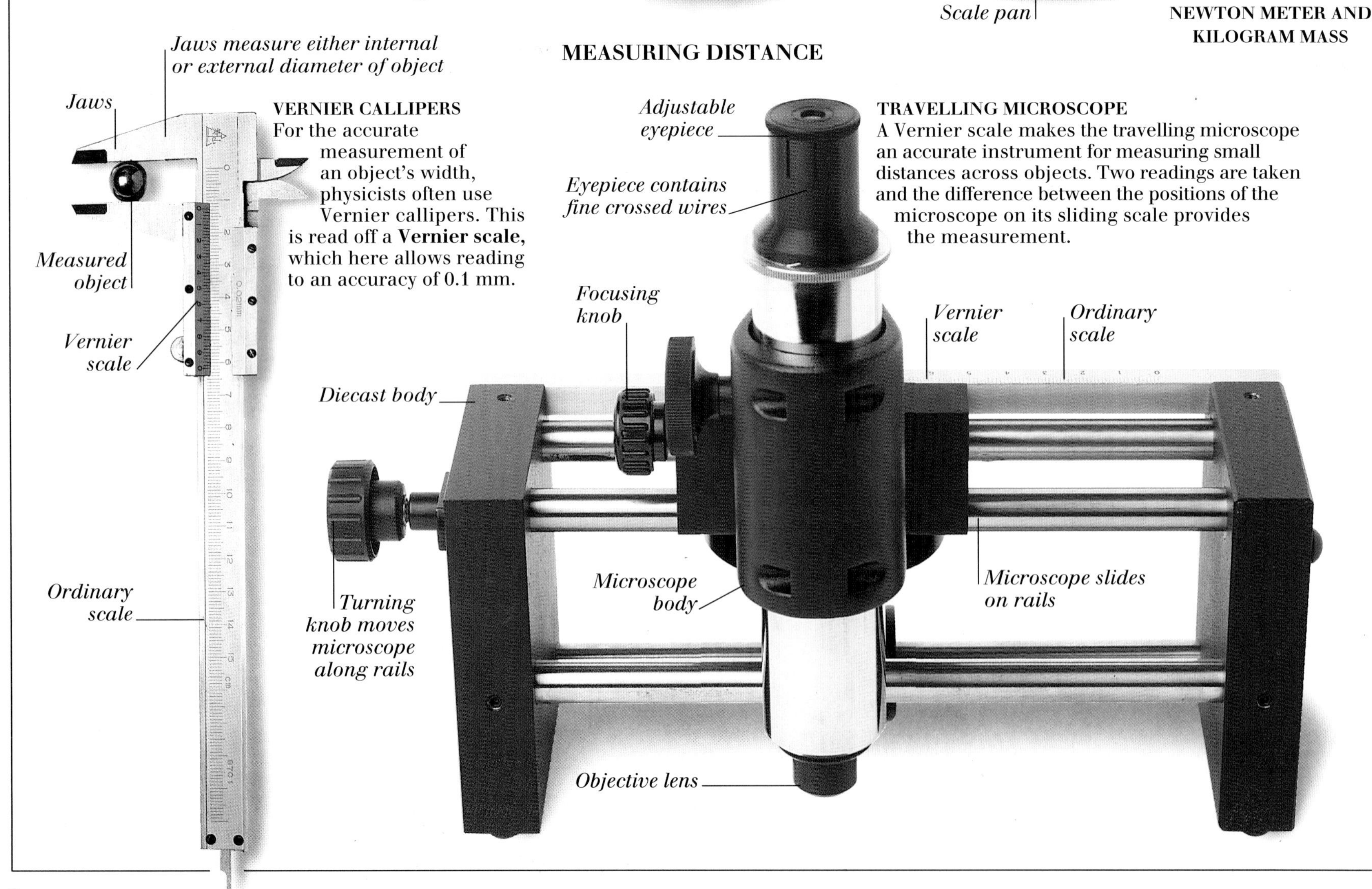

VERNIER CALLIPERS

For the accurate measurement of an object's width, physicists often use Vernier callipers. This is read off a **Vernier scale,** which here allows reading to an accuracy of 0.1 mm.

TRAVELLING MICROSCOPE

A Vernier scale makes the travelling microscope an accurate instrument for measuring small distances across objects. Two readings are taken and the difference between the positions of the microscope on its sliding scale provides the measurement.

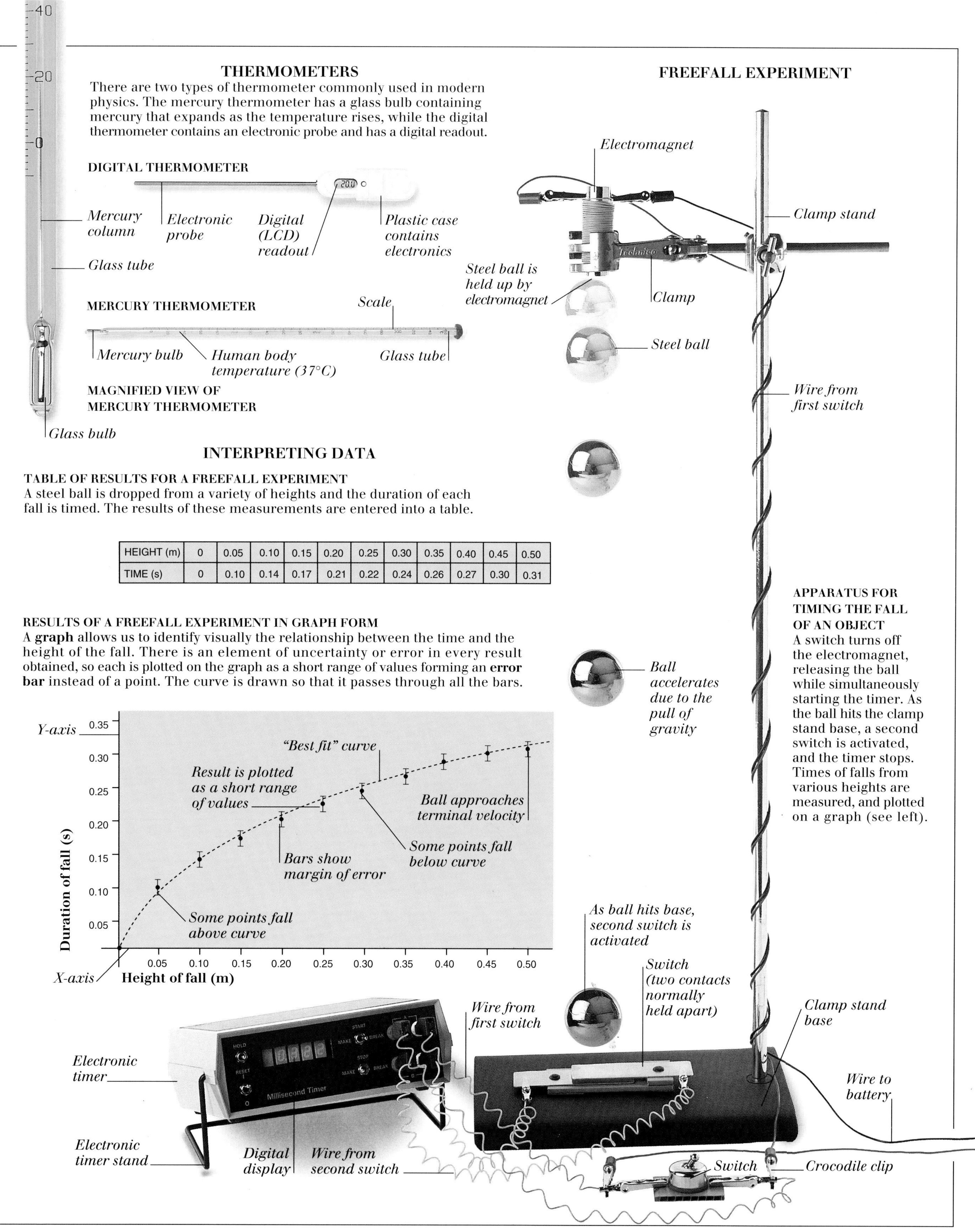

THERMOMETERS

There are two types of thermometer commonly used in modern physics. The mercury thermometer has a glass bulb containing mercury that expands as the temperature rises, while the digital thermometer contains an electronic probe and has a digital readout.

INTERPRETING DATA

TABLE OF RESULTS FOR A FREEFALL EXPERIMENT

A steel ball is dropped from a variety of heights and the duration of each fall is timed. The results of these measurements are entered into a table.

HEIGHT (m)	0	0.05	0.10	0.15	0.20	0.25	0.30	0.35	0.40	0.45	0.50
TIME (s)	0	0.10	0.14	0.17	0.21	0.22	0.24	0.26	0.27	0.30	0.31

RESULTS OF A FREEFALL EXPERIMENT IN GRAPH FORM

A **graph** allows us to identify visually the relationship between the time and the height of the fall. There is an element of uncertainty or error in every result obtained, so each is plotted on the graph as a short range of values forming an **error bar** instead of a point. The curve is drawn so that it passes through all the bars.

FREEFALL EXPERIMENT

APPARATUS FOR TIMING THE FALL OF AN OBJECT

A switch turns off the electromagnet, releasing the ball while simultaneously starting the timer. As the ball hits the clamp stand base, a second switch is activated, and the timer stops. Times of falls from various heights are measured, and plotted on a graph (see left).

Forces 1

A FORCE IS A PUSH OR PULL. Forces can be large or small – the usual unit of force is the newton (N), and can be measured using a **newton meter** (see pp. 8-9). Force can be applied to objects at a distance or by making contact. **Gravity** (see pp. 12-13) and **electromagnetism** (see pp. 34-35) are examples of forces that can act at a distance. When more than one force acts on an object, the combined force is called the **resultant**. The resultant of several forces depends on their size and direction. The object is in **equilibrium** if the forces on an object are balanced with no overall resultant. An object on a solid flat surface will be in equilibrium, because the surface produces a **reaction** force to balance the object's **weight**. If the surface slopes, the object's weight is no longer completely cancelled by the reaction force and part of the weight, called a **component**, remains, pulling the object towards the bottom of the slope. Forces can cause rotation as well as straight line motion. If an object is free to rotate about a certain point, then a force can have a turning effect, known as a **moment**.

REACTION FORCES

FORCES ON A LEVEL SURFACE

A table provides a force called a reaction, which exactly balances the weight of an object placed upon it. The resultant force is zero, so the object does not fall through the table.

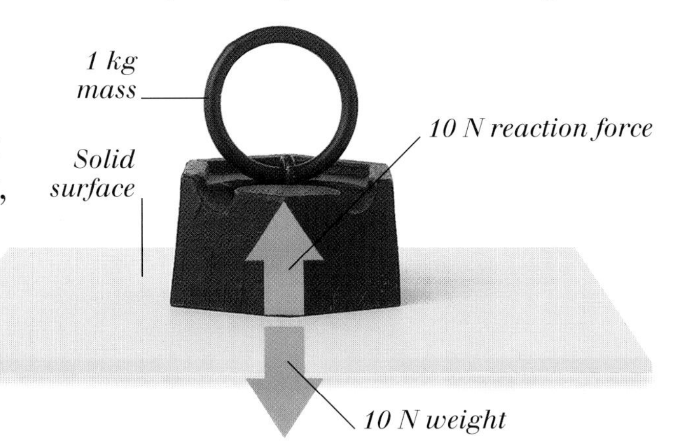

FORCES ON A SHALLOW SLOPE

Gravity acts downwards on the 1 kg mass shown. The slope provides a reaction force that acts upwards, perpendicular to the slope and counteracts some of the weight. All that remains of the weight is a force acting down the slope.

FORCES ON A STEEP SLOPE

As the slope is made steeper, the reaction force of the slope decreases, and the force pulling the mass down the slope – which is measured by the newton meter – increases. This force can pull objects downhill.

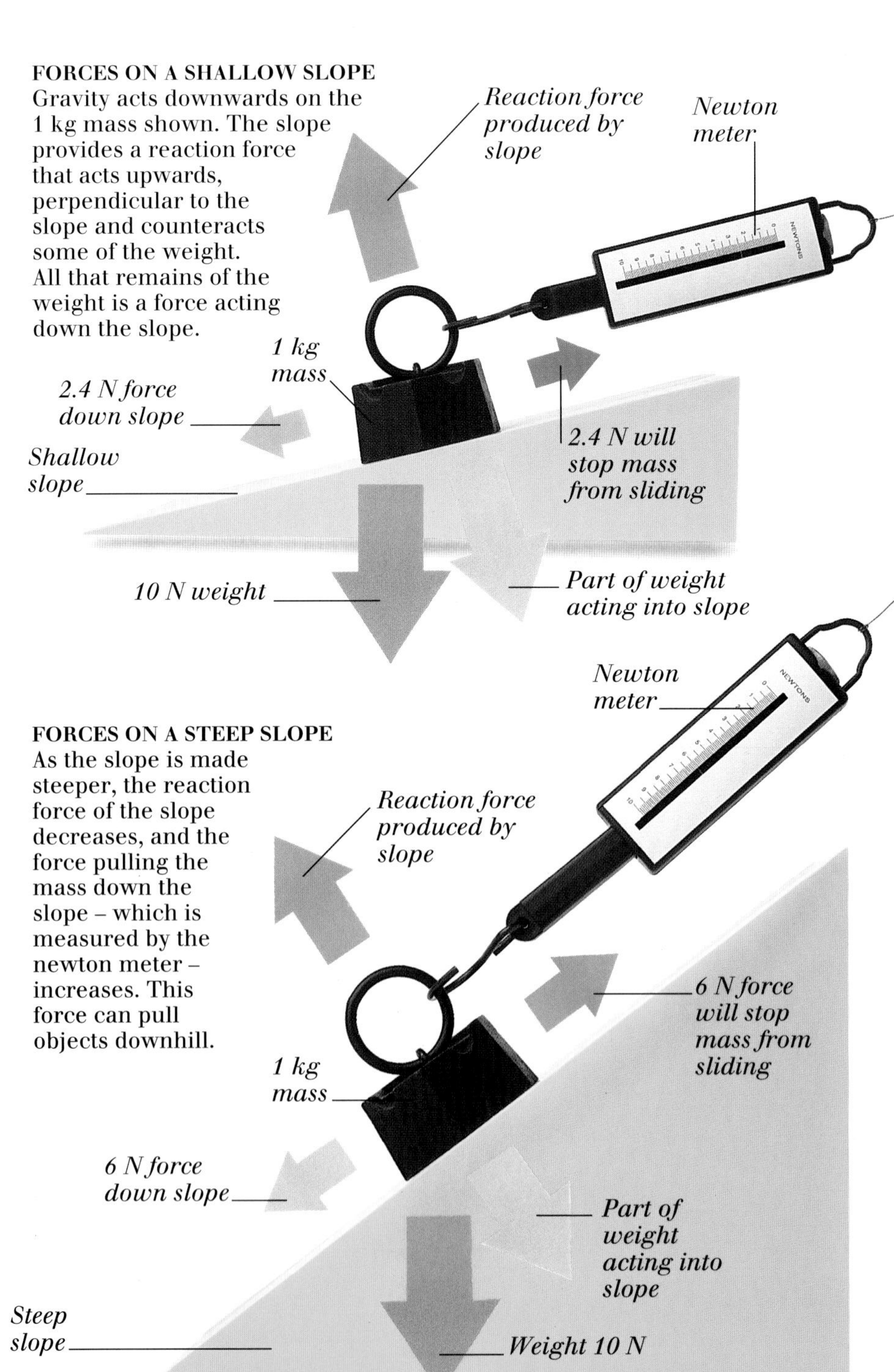

RESULTANT FORCE

A 1 kg **mass** has a weight of 10 N. Here, this weight is supported by two lengths of wire. Each wire carries a force that pulls against the other at an angle. The combination or resultant of these forces is 10 N vertically upwards and exactly balances the weight. The force carried by each wire is measured by newton meters.

Wire

Newton meters held at an angle

Resultant upward force of 10 N exactly balances 10 N weight

Reading 5.8 N

Reading 5.8 N

Force acts at an angle

Force acts at an angle

1 kg mass

10 N weight

THE METER READINGS

Between them, the two wires support a weight of 10 N, so why is the reading on each newton meter more than 5 N? As well as pulling upwards, the wires are pulling sideways against each other, so the overall force showing on each meter is 5.8 N.

TURNING FORCES

TURNING FORCES AROUND A PIVOT

A force acting on an object that is free to rotate will have a turning effect, or turning force, also known as a moment. The moment of a force is equal to the size of the force multiplied by the distance of the force from the turning point around which it acts (see p. 54). It is measured in newton metres (Nm). The mass below exerts a weight of 10 N downwards on a pivoted beam. The newton meter – twice as far from the pivot – measures 5 N, the upward force needed to stop the beam turning. The clockwise moment created by the weight and anticlockwise moment created by the upward pull on the newton meter are equal, and the object is therefore in equilibrium.

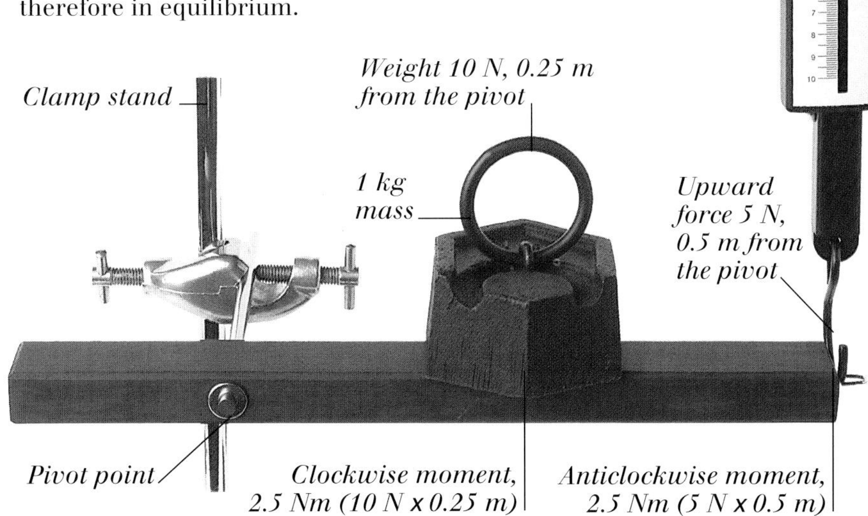

OBJECT SUSPENDED AT CENTRE OF GRAVITY

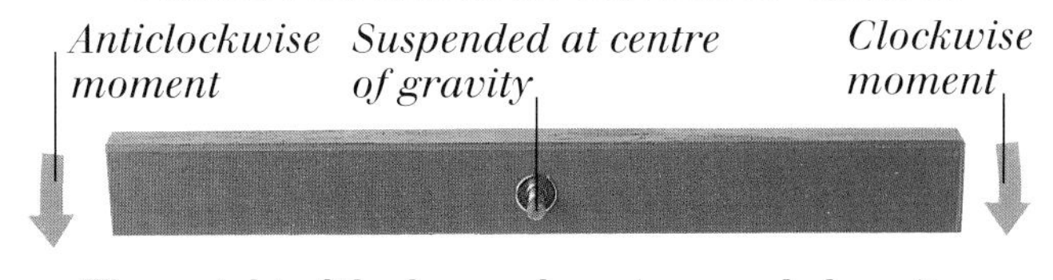

The weight of the beam above is spread along its length. The moments are balanced if the object is suspended at its centre of gravity.

OBJECT SUSPENDED AWAY FROM CENTRE OF GRAVITY

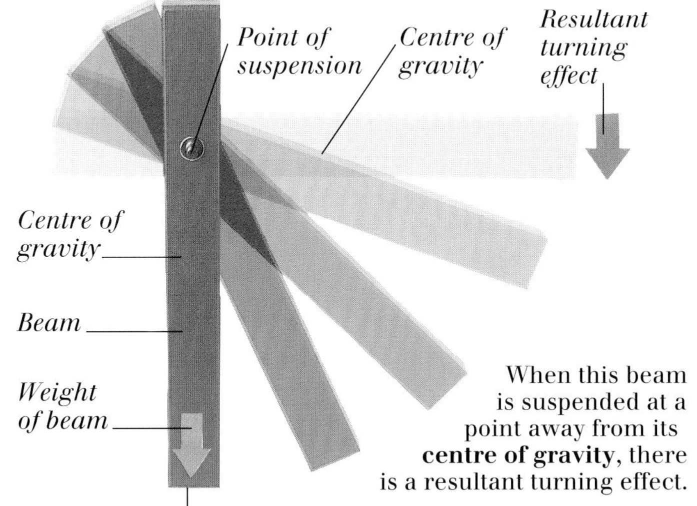

When this beam is suspended at a point away from its **centre of gravity**, there is a resultant turning effect.

PRESSURE

Why can a drawing pin be pushed into a wall, and yet a building will not sink into the ground? Forces can act over large or small areas. A force acting over a large area will exert less **pressure** than the same force acting over a small area. The pressure exerted on an area can be worked out simply by dividing the applied force by the area over which it acts (see p. 54). Pressure is normally measured in units of newtons per square metre (Nm^{-2}). A drawing pin concentrates force to produce high pressure, whereas the foundations of a building spread the load to reduce pressure. Gases also exert pressure (see pp. 28-29).

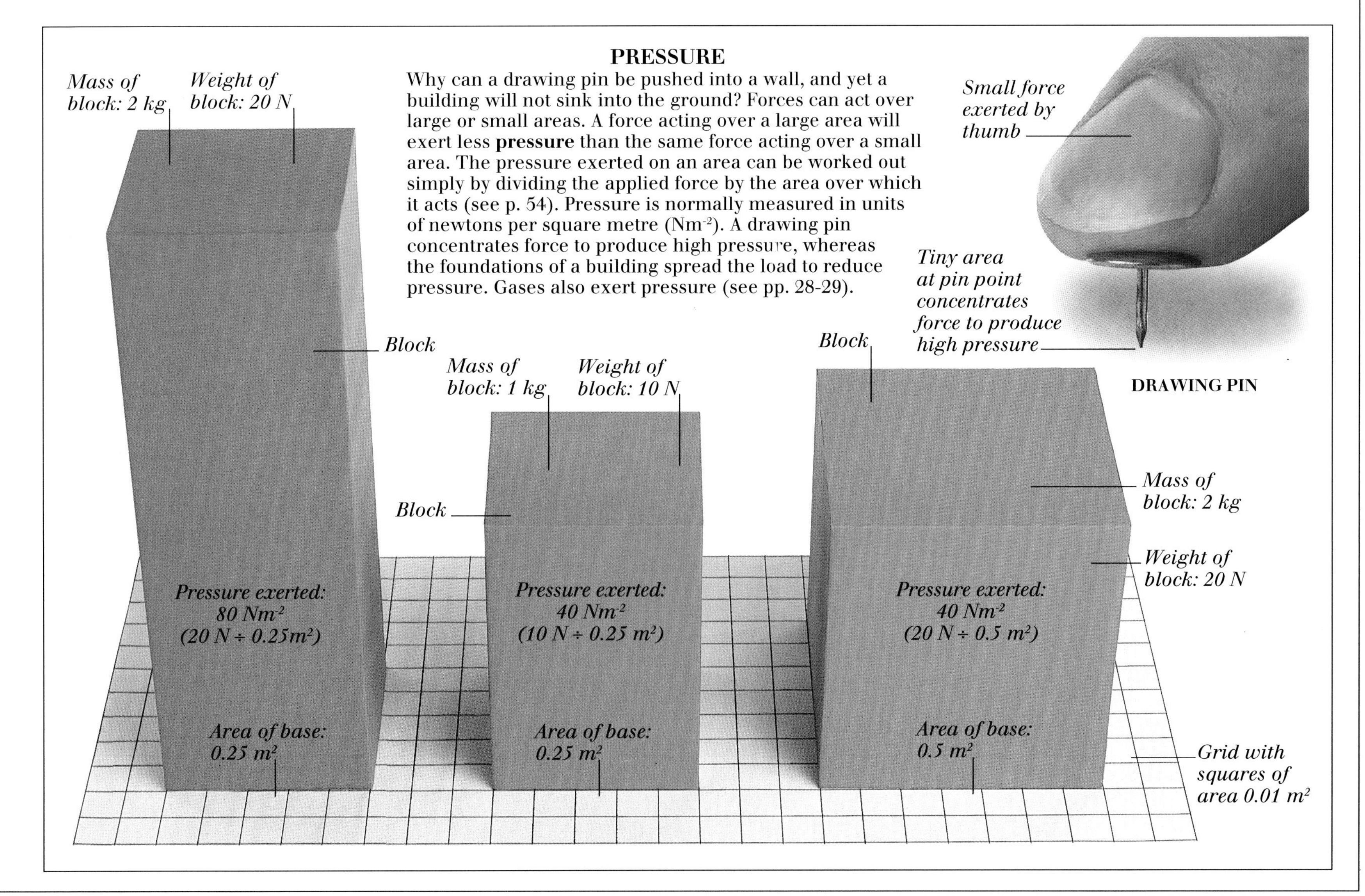

Forces 2

WHEN THE FORCES ON AN OBJECT do not cancel each other out, they will change the motion of the object. The object's speed, direction of motion, or both will change. The rules governing the way forces change the motion of objects were first worked out by Sir Isaac Newton. They have become known as Newton's Laws. The greater the mass of an object, the greater the force needed to change its motion. This resistance to change in motion is called **inertia**. The **speed** of an object is usually measured in metres per second (ms^{-1}). **Velocity** is the speed of an object in a particular direction. **Acceleration**, which only occurs when a force is applied, is the rate of change in speed. It is measured in metres per second per second, or metres per second squared (ms^{-2}). One particular force keeps the Moon in orbit around the Earth and the Earth in orbit around the Sun. This is the force of **gravity** or **gravitation**; its effects can be felt over great distances.

NEWTON'S SECOND LAW IN ACTION

Trucks have a greater mass than cars. According to Newton's second law (see right) a large mass requires a larger force to produce a given acceleration. This is why a truck needs to have a larger engine than a car.

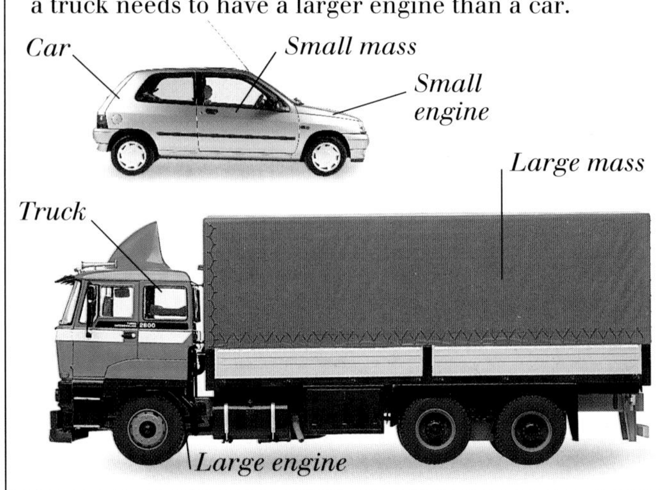

NEWTON'S LAWS

NEWTON'S FIRST LAW

When no force acts on an object, it will remain in a state of rest or continue its uniform motion in a straight line.

No force, no acceleration: state of rest

No force, no acceleration: uniform motion

NEWTON'S SECOND LAW

When a force acts on an object, the motion of the object will change. This change in motion is called acceleration and is equal to the size of the force divided by the mass of the object on which it acts (see p. 54).

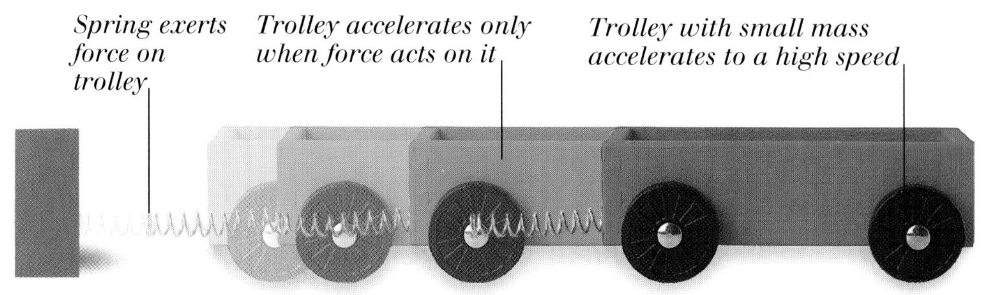

Force acts on small mass: large acceleration

Same force acts on large mass: small acceleration

NEWTON'S THIRD LAW

If one object exerts a force on another, an equal and opposite force, called the reaction force, is applied by the second to the first.

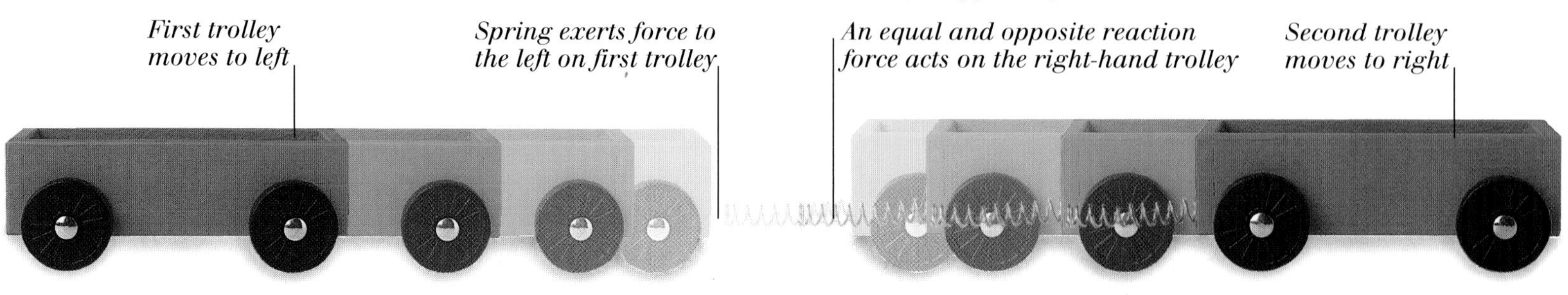

Action and reaction

FORCE AND MOTION

In the images below, each row of balls is a record of the motion of one ball, photographed once each second beside a ruler. This shows how far the ball moved during that second and each subsequent second, giving a visual representation of speed and acceleration.

SPEED

Speed is the distance an object travels in a set amount of time. It is calculated by dividing distance covered by time taken (see p. 54). In physics, speed is measured in metres per second (ms^{-1}).

MOMENTUM

The momentum of an object is equal to its mass multiplied by its velocity (see p. 54). Momentum is measured in kilogram metres per second ($kgms^{-1}$). The two balls below have the same momentum.

ACCELERATION

Acceleration is the rate that the speed of an object changes. It is calculated by dividing the change in speed by the time it took for that change (see p. 54). It is measured in metres per second per second (ms^{-2}).

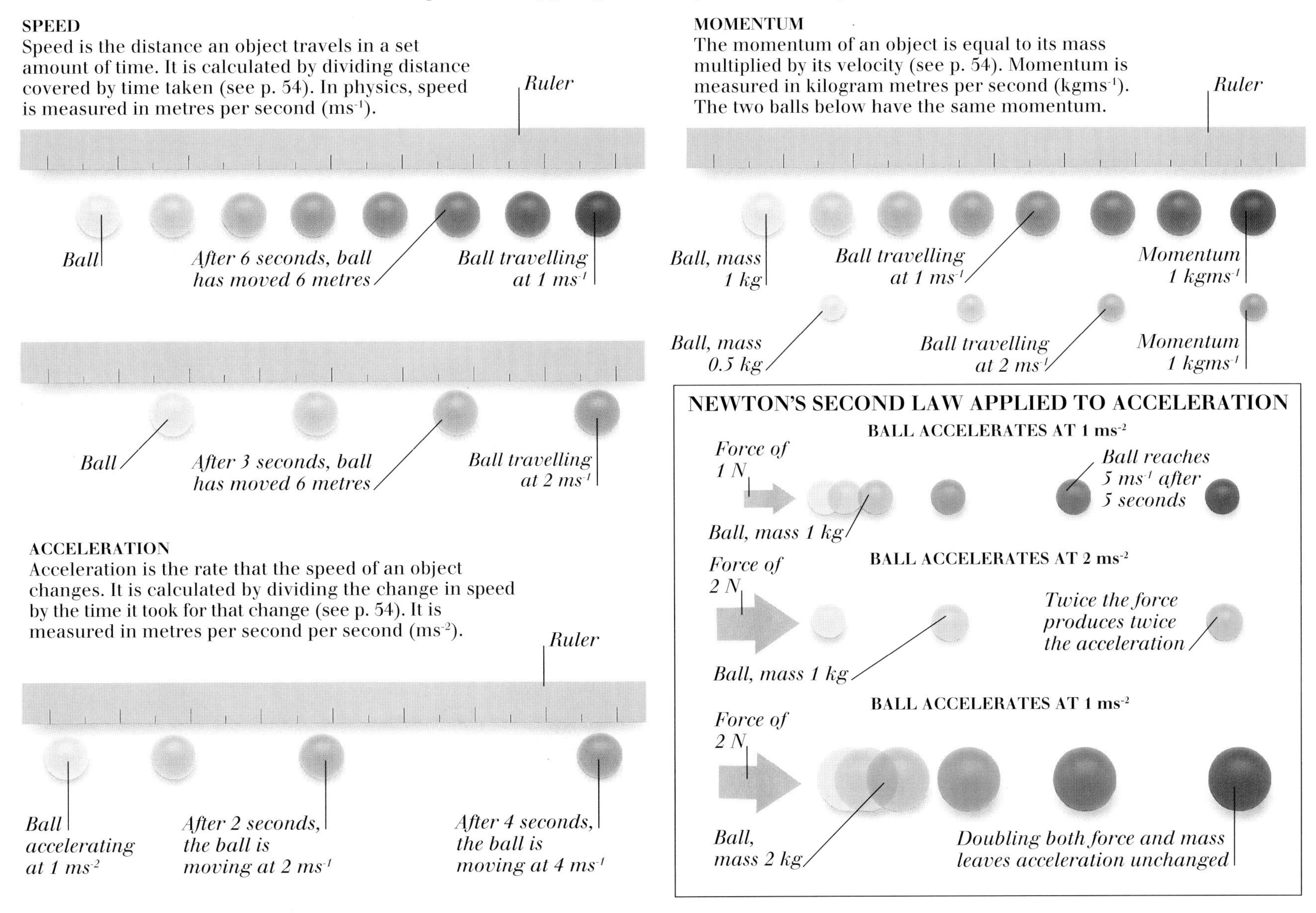

GRAVITATIONAL FORCE

Gravitation, or gravity, is a force that acts on all matter. The force between any two objects depends upon their masses and the distance between them (see p. 54).

If the Moon had twice the mass that it does, the force between the Earth and Moon would be twice as large.

If the Moon were half the distance from the Earth, the gravitational force would be four times as large. This is because the force depends upon the distance squared.

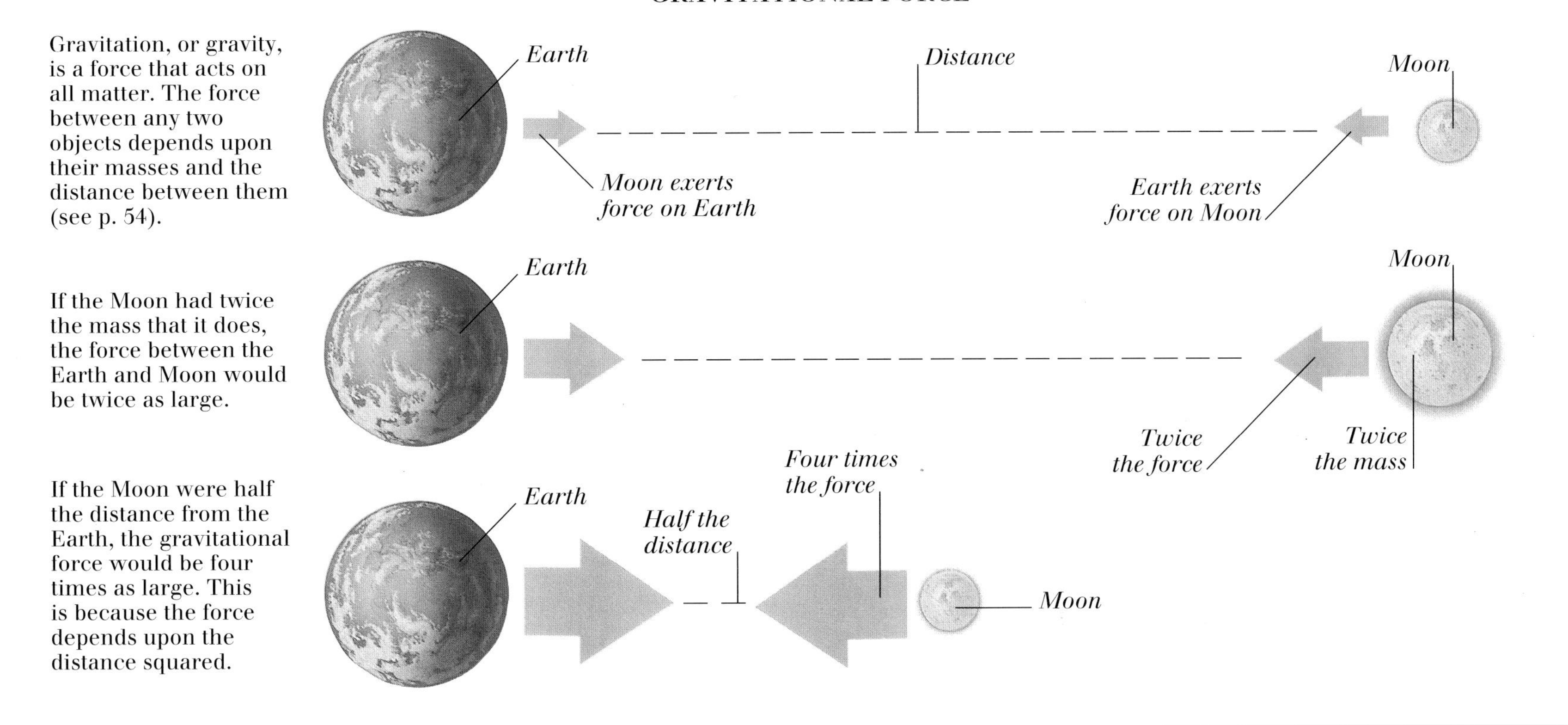

Friction

FRICTION IS A FORCE THAT SLOWS DOWN or prevents motion. A familiar form of friction is air resistance, which limits the speed at which objects can move through the air. Between touching surfaces, the amount of friction depends on the nature of the surfaces and the force or forces pushing them together. It is the joining or bonding of the **atoms** at each of the surfaces that causes the friction. When you try to pull an object along a table, the object will not move until the **limiting friction** supplied by these bonds has been overcome. Friction can be reduced in two main ways: by lubrication or by the use of rollers. Lubrication involves the presence of a **fluid** between two surfaces; fluid keeps the surfaces apart, allowing them to move smoothly past one another. Rollers actually use friction to grip the surfaces and produce rotation. Instead of sliding against one another, the surfaces produce turning forces, which cause each roller to roll. This leaves very little friction to oppose motion.

AIR RESISTANCE

Air resistance is a type of friction that occurs when an object moves through the air. The faster an object moves, the greater the air resistance. Falling objects accelerate to a speed called **terminal velocity**, at which the air resistance exactly balances the object's weight. At this speed, there is no **resultant force** and so no further acceleration can occur.

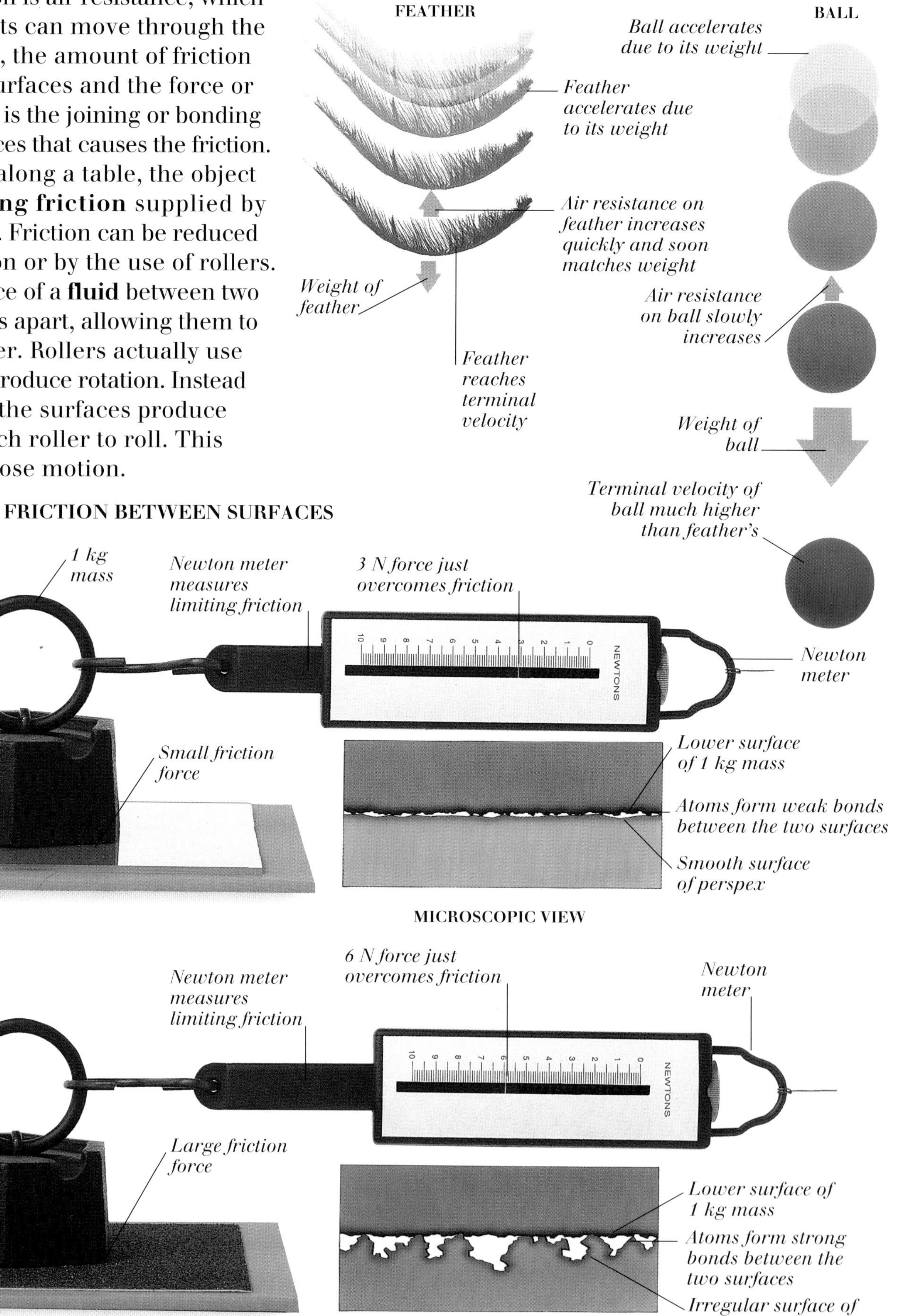

LOW LIMITING FRICTION
Limiting friction must be overcome before surfaces can move over each other. Smooth surfaces produce little friction. Only a small amount of force is needed to break the bonds between atoms.

HIGH LIMITING FRICTION
Rougher surfaces produce a larger friction force. Stronger bonds are made between the two surfaces and more energy is needed to break them. The mass requires a large force to slide over sandpaper.

MOTORCYCLE BRAKE

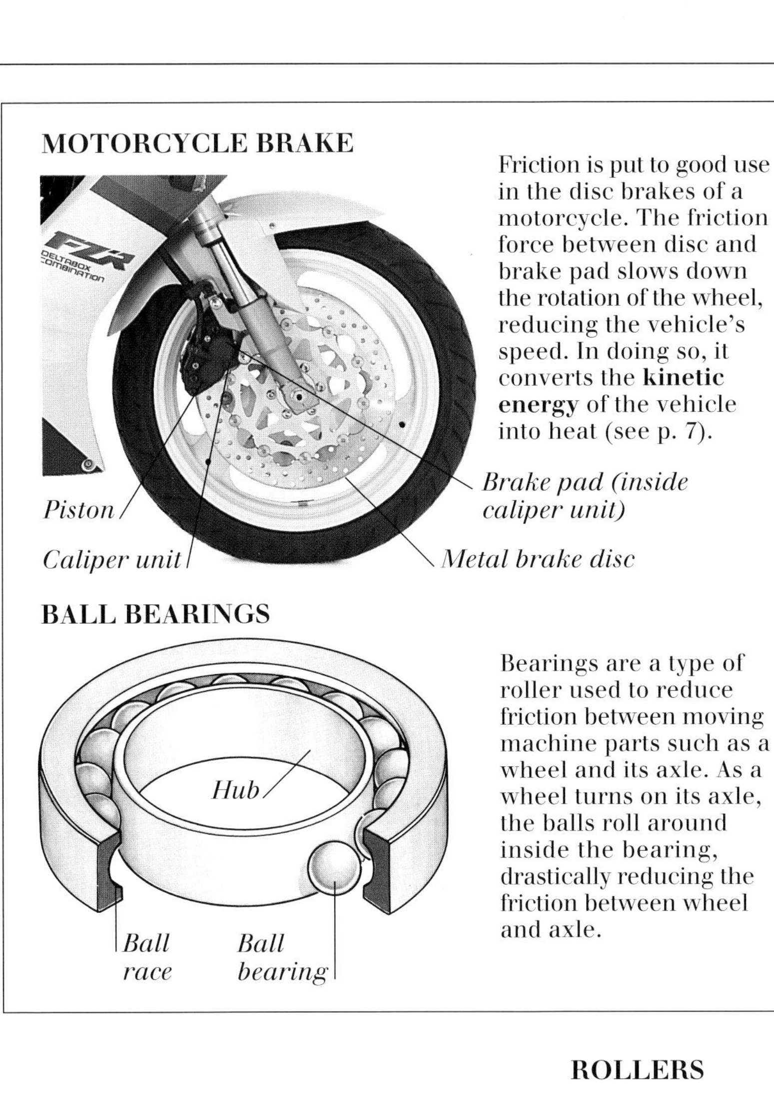

Friction is put to good use in the disc brakes of a motorcycle. The friction force between disc and brake pad slows down the rotation of the wheel, reducing the vehicle's speed. In doing so, it converts the **kinetic energy** of the vehicle into heat (see p. 7).

BALL BEARINGS

Bearings are a type of roller used to reduce friction between moving machine parts such as a wheel and its axle. As a wheel turns on its axle, the balls roll around inside the bearing, drastically reducing the friction between wheel and axle.

LUBRICATION

The presence of oil or another fluid between two surfaces keeps the surfaces apart. Because fluids (liquids or gases) flow, they allow movement between surfaces. Here, a lubricated kilogram mass slides down a slope, while an unlubricated one is prevented from moving by friction.

Unlubricated mass remains stationary

Patch of oil reduces friction

High friction prevents mass from moving

Inclined plane

1 kg mass

Lubricated mass moves down slope

ROLLERS

THE ACTION OF A ROLLER ON A SLOPE
Friction causes the roller to grip the slope so that it turns. If there were no friction, the roller would simply slide down the slope.

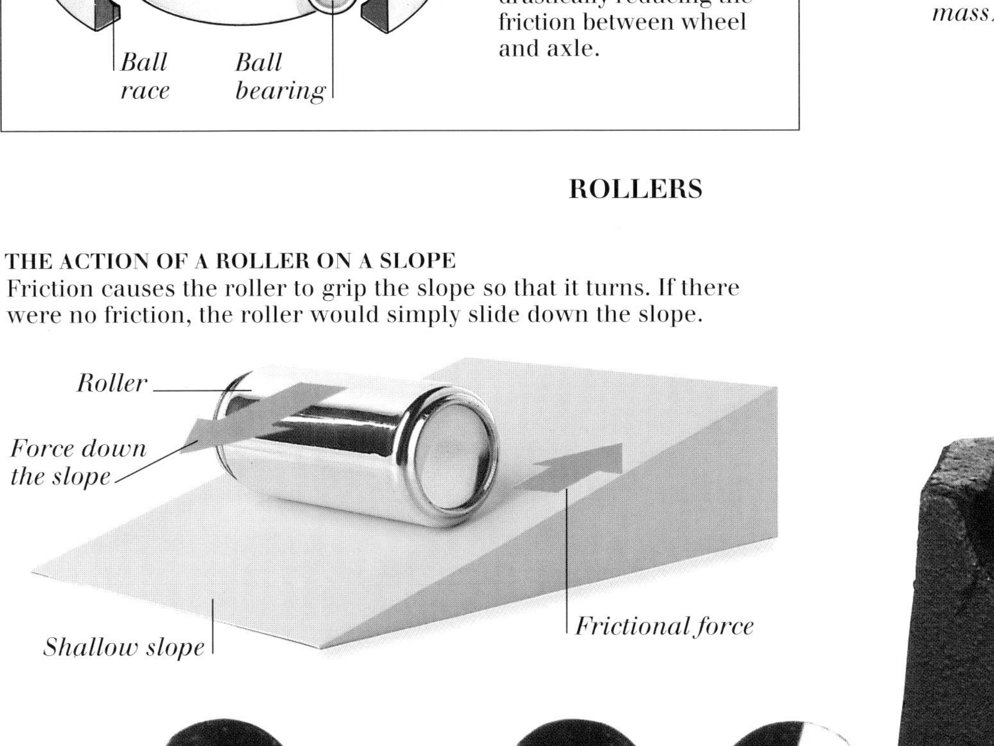

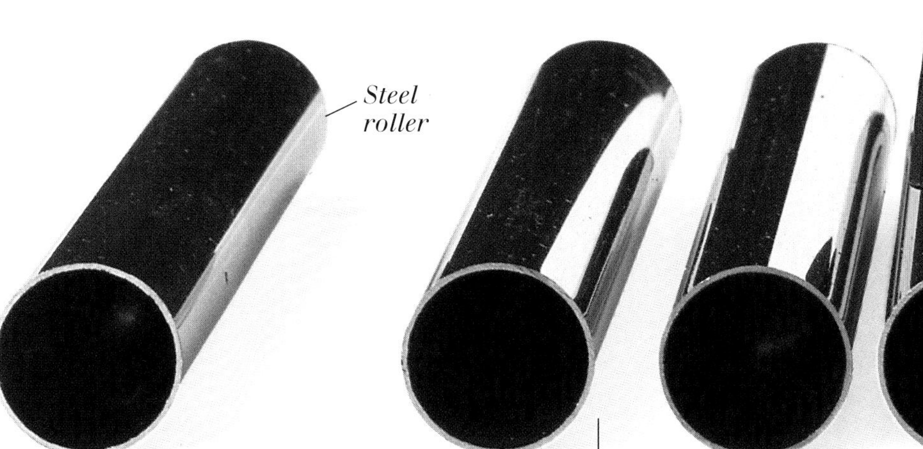

1 kg mass

USING ROLLERS TO AVOID FRICTION
Rollers placed between two surfaces keep the surfaces apart. The rollers allow the underside of the kilogram mass to move freely over the ground. An object placed on rollers will move smoothly if pushed or pulled.

Friction forces between surfaces create a turning force that turns the rollers

Mass moves smoothly over surface

Underside of 1 kg mass

Simple machines

IN PHYSICS, A MACHINE IS ANY DEVICE that can be used to transmit a **force** (see pp. 10-11) and, in doing so, change its size or direction. When using a simple pulley, a type of machine, a person can lift a load by pulling downwards on the rope. By using several pulleys connected together as a block and tackle, the size of the force can be changed too, so that a heavy load can be lifted using a small force. Other simple machines include the inclined plane, the lever, the screw, and the wheel and axle. All of these machines illustrate the concept of **work**. Work is the amount of energy expended when a force is moved through a distance. The force applied to a machine is called the effort, while the force it overcomes is called the load. The effort is often smaller than the load, for a small effort can overcome a heavy load if the effort is moved through a larger distance. The machine is then said to give a mechanical advantage. Although the effort will be smaller when using a machine, the amount of work done, or energy used, will be equal to or greater than that without the machine.

WEDGE

The axe is a wedge. The applied force moves a long way into the wood, producing a larger force, which pushes the wood apart a short distance.

Axe handle

Metal axe blade

Small force applied

Block of wood

Large force produced

Wood splits apart

AN INCLINED PLANE

The force needed to drag an object up a slope is less than that needed to lift it vertically. However, the distance moved by the object is greater when pulled up the slope than if it were lifted vertically.

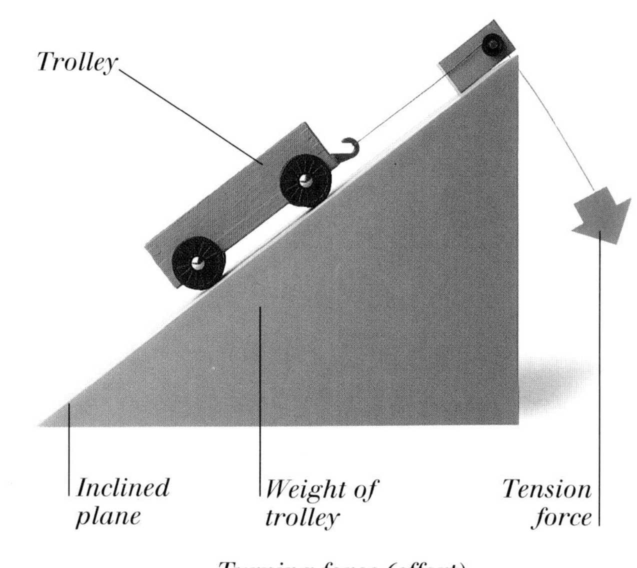

SCREW

A screw is like an inclined plane wrapped around a shaft. The force that turns the screw is converted to a larger one, which moves a shorter distance and drives the screw in.

Screw thread unravelled

Screw thread

Screw is pulled into wood with force greater than the effort

CORKSCREW

The corkscrew is a clever combination of several different machines. The screw pulls its way into the cork, turned by a wheel and axle. The cork is lifted by a pair of class one levers (see opposite).

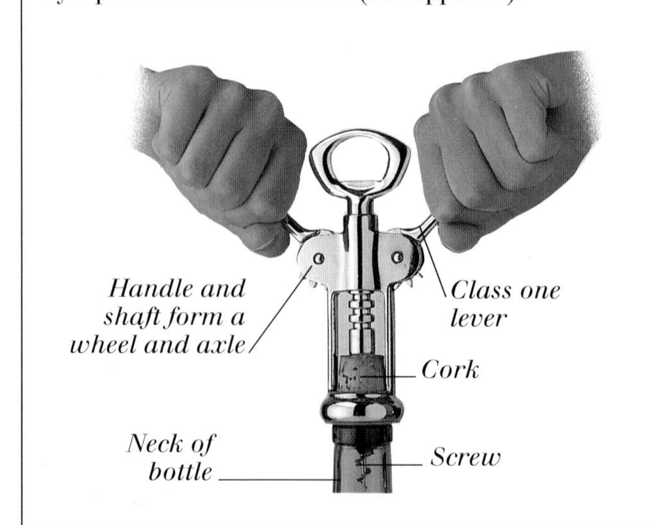

PULLEYS

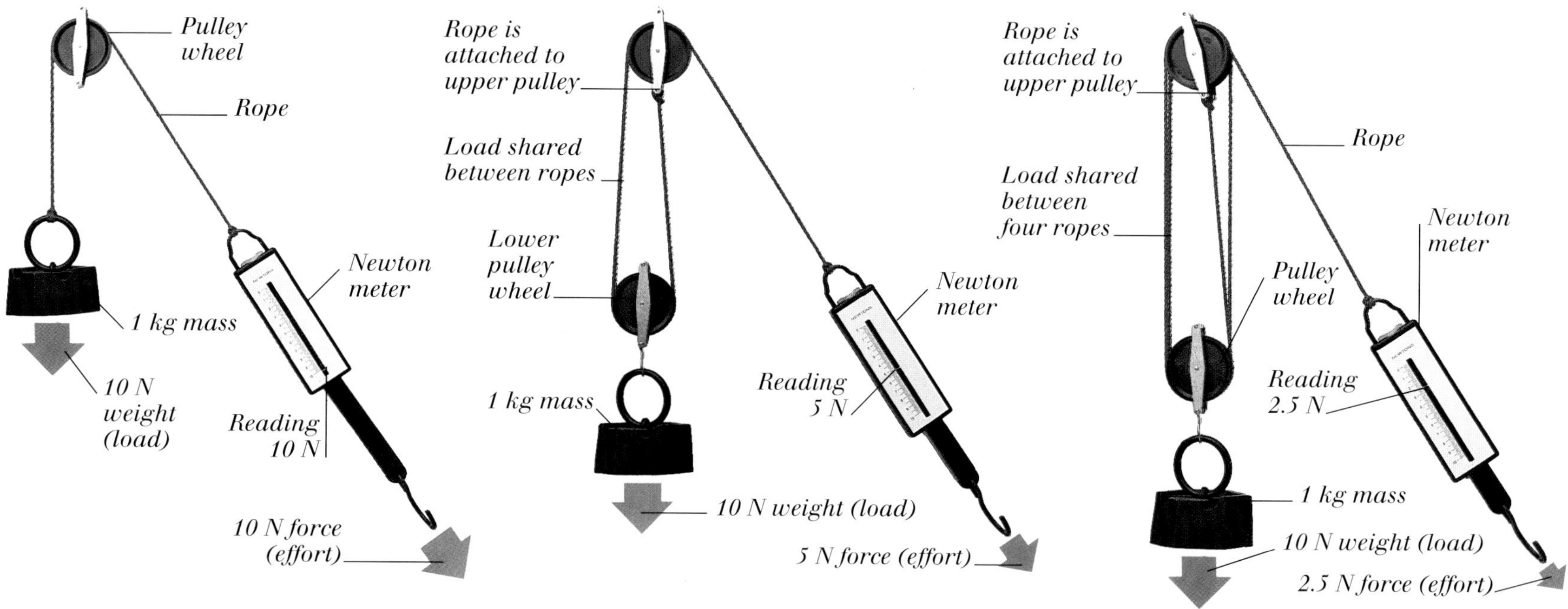

SIMPLE PULLEY
A simple pulley changes the direction of a force but not its size. Here a one kg mass, weighing ten newtons is lifted by a ten newton force. The mass and the other end of the rope move through the same distance.

DOUBLE PULLEY
A double pulley will lift a one kg mass with only a five newton effort, because the force in the rope doubles up as the rope does. However, the rope must be pulled twice as far as the load is lifted.

QUADRUPLE PULLEY
Lifting a one kg mass with a quadruple pulley, in which the rope goes over four pulley wheels, feels almost effortless. However, the rope must be pulled four times as far as the load is lifted.

THREE CLASSES OF LEVER

CLASS ONE LEVER
In a class one lever, the **fulcrum** (pivot point) is between the effort and the load. The load is larger than the effort, but it moves through a smaller distance.

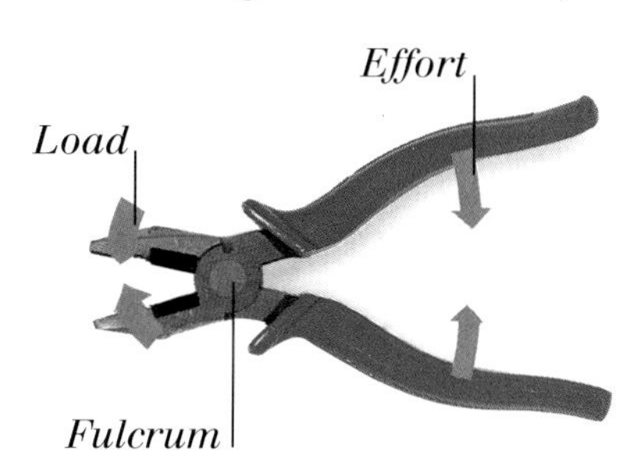

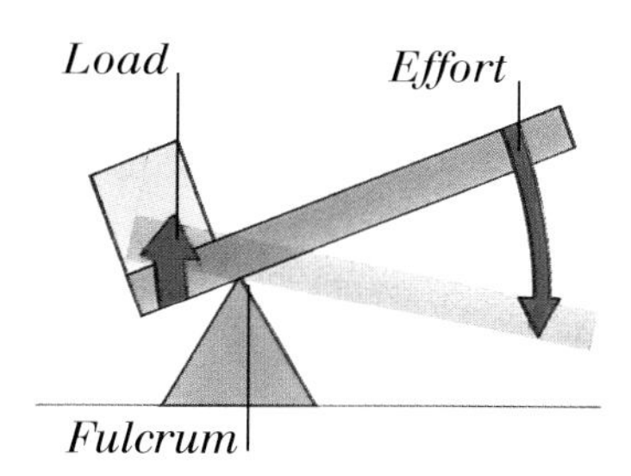

CLASS TWO LEVER
In a class two lever, the load is between the fulcrum and effort. Here again, the load is greater than the effort and it moves through a smaller distance.

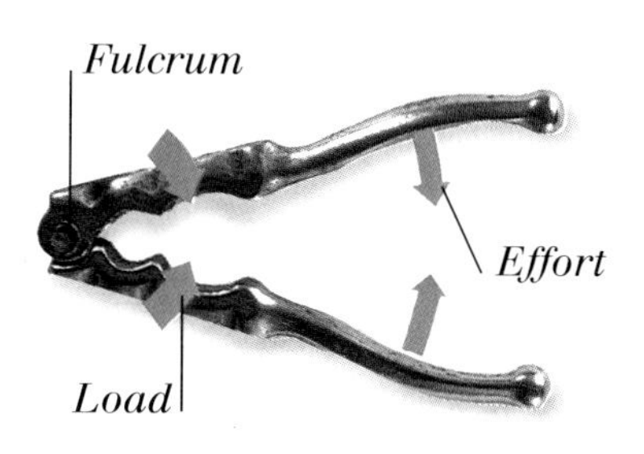

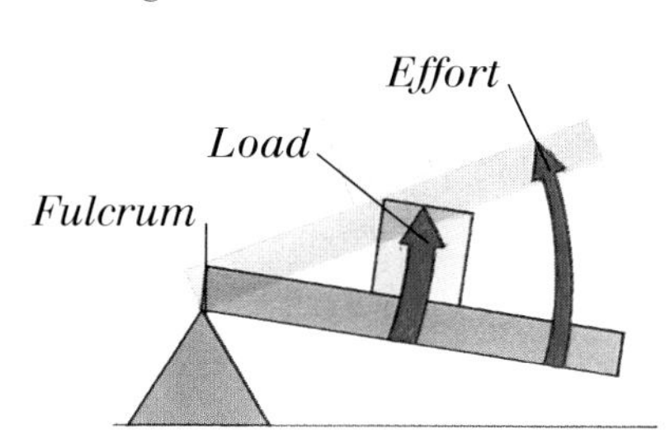

CLASS THREE LEVER
In a class three lever, the effort is between the fulcrum and the load. In this case, the load is less than the effort but it moves through a greater distance.

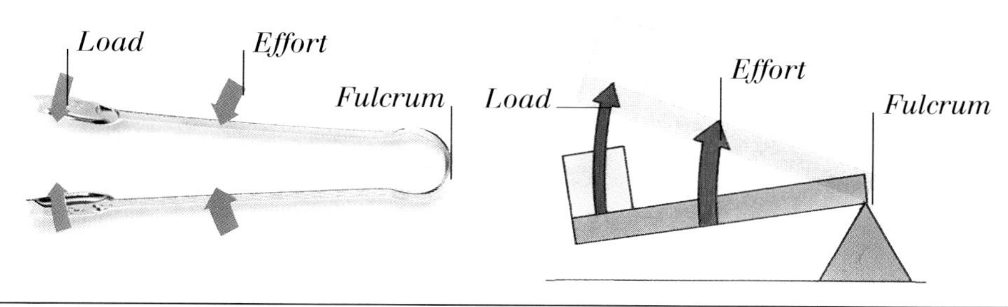

WHEEL AND AXLE

As the pedal and chainwheel of a bicycle turn through one revolution, the pedal moves farther than the links of the chain. For this reason, the force applied to the chain is greater than the force applied to the pedal. The steering wheel of a car is another example of a wheel and axle.

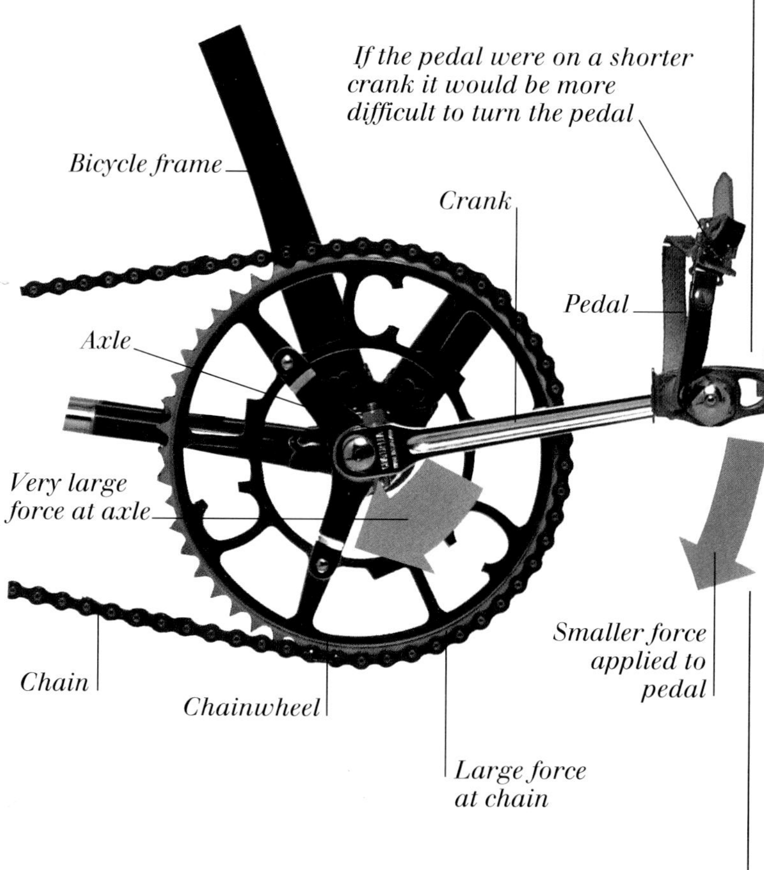

Circular motion

WHEN AN OBJECT MOVES IN A CIRCLE, its direction is continuously changing. Any change in direction requires a **force** (see pp. 12-13). The force required to maintain circular motion is called **centripetal force**. The size of this force depends on the size of the circle, and the mass and speed of the object (see p. 54). The centripetal force that keeps an object whirling round on the end of a string is caused by **tension** (see pp. 24-25) in the string. When the centripetal force ceases – for example, if the string breaks – the object flies off in a straight line, since no force is acting upon it. **Gravity** (see pp. 10-11) is the centripetal force that keeps planets such as the Earth in orbit. Without this centripetal force, the Earth would move in a straight line through space. On a smaller scale, without friction to provide centripetal force, a motorcyclist could not steer around a bend. Spinning, a form of circular motion, gives **gyroscopes** stability.

CENTRIPETAL FORCE

In the experiment below, centripetal force is provided by tension in a length of string, which keeps a 1 kg mass moving in a circle. The mass can move freely as it floats like a hovercraft on the jets of air supplied from beneath it. When the circle is twice as large, half the force is needed. However, moving twice as fast requires four times the force (see p. 54).

CONTROL EXPERIMENT

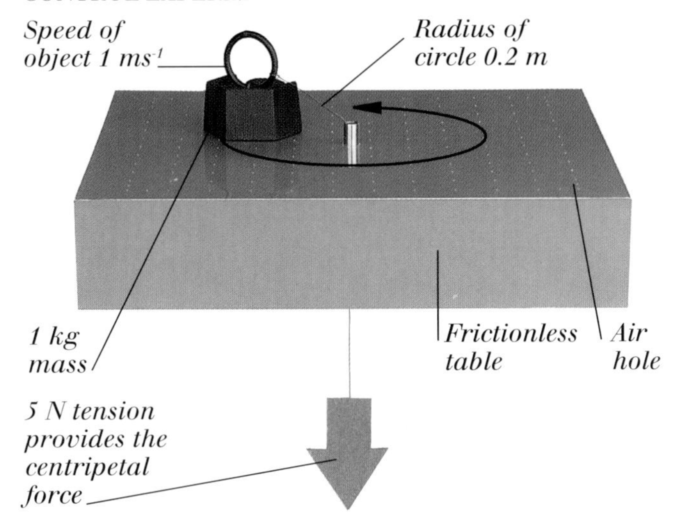

TWICE THE SPEED, FOUR TIMES THE FORCE

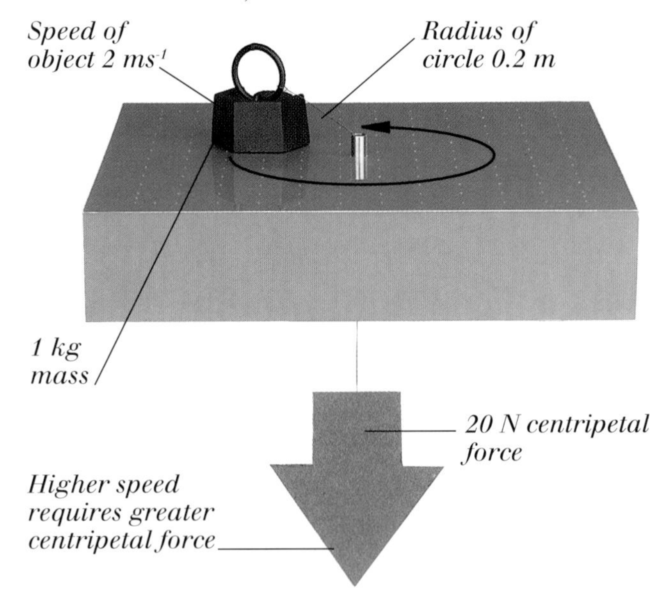

TWICE THE RADIUS, HALF THE FORCE

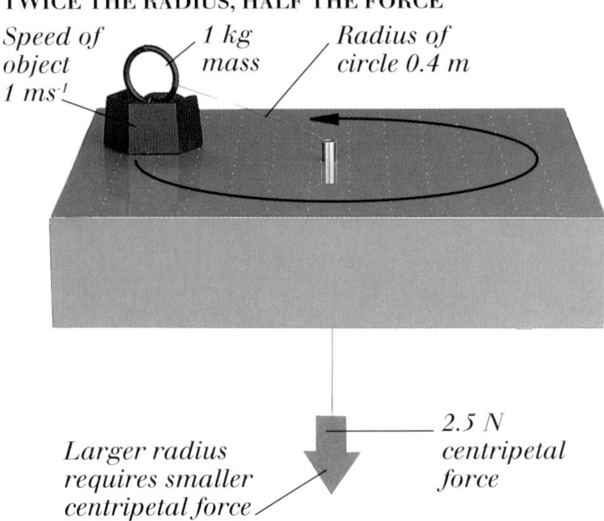

MOTION IN A CIRCLE

ASPECTS OF CIRCULAR MOTION

The force that continuously changes the direction of an object moving in a circle is called centripetal force. It is directed towards the centre of the circle. The smaller the radius of the circle, the larger the force needed.

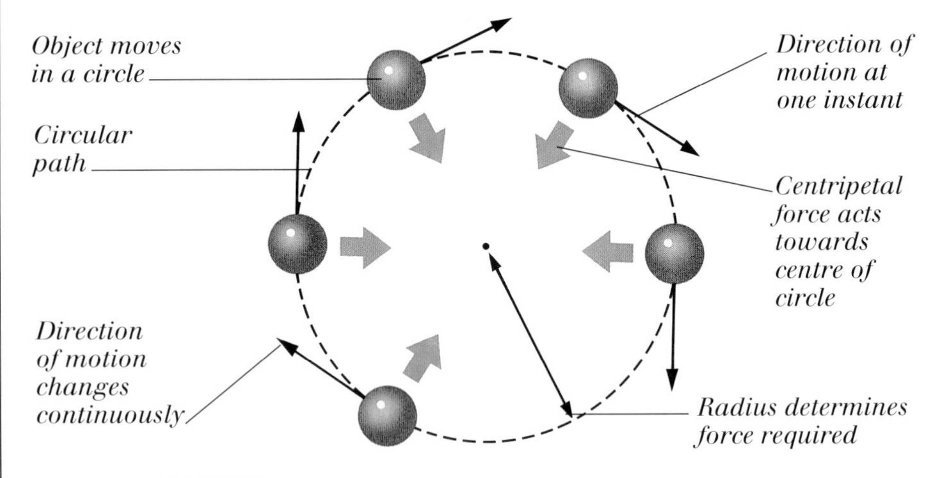

HAMMER THROWER

Tension in muscles provides the centripetal force needed to whirl a hammer round in a circle. When the thrower releases the chain, no force acts upon the hammer and it moves off in a straight line.

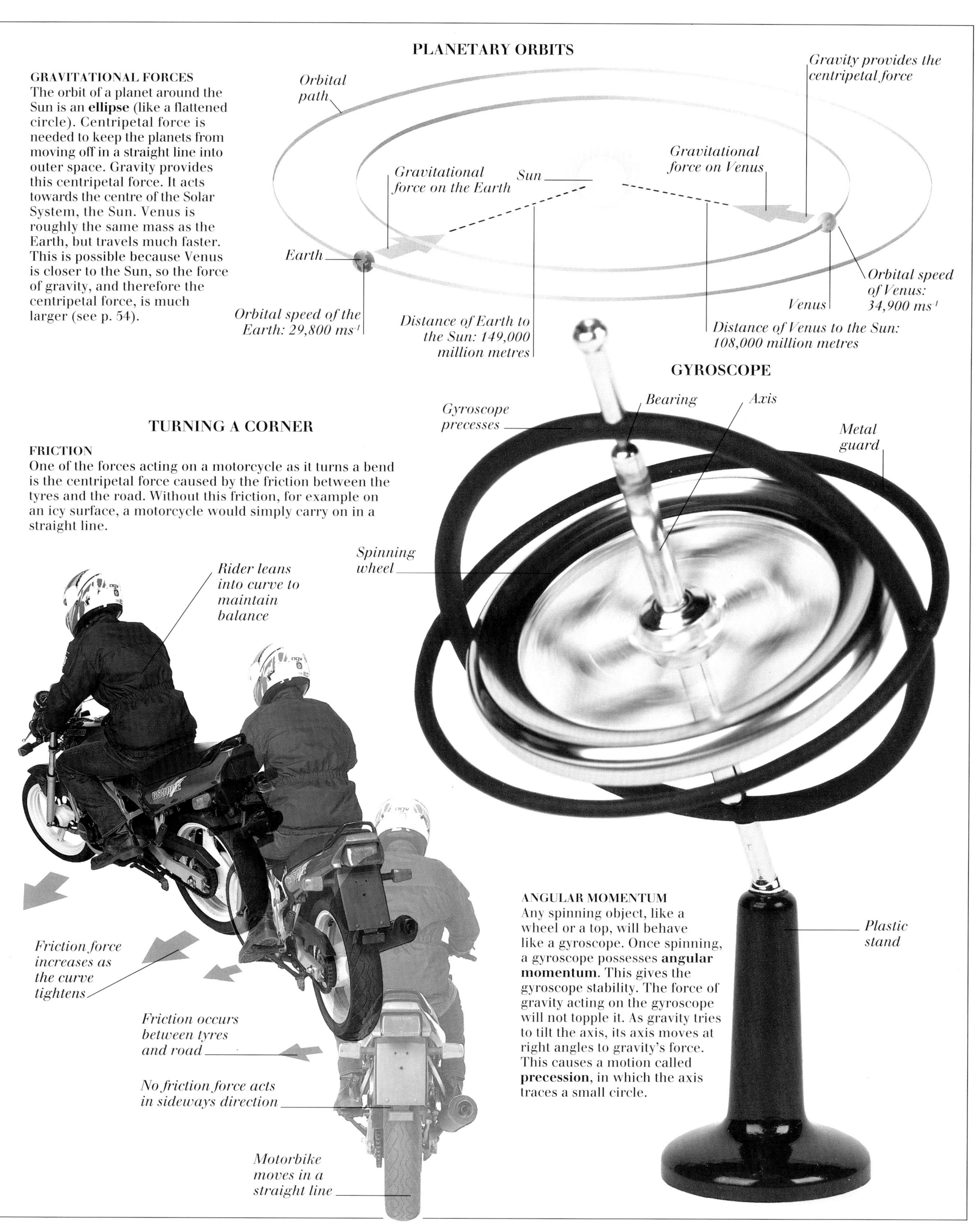

PLANETARY ORBITS

GRAVITATIONAL FORCES

The orbit of a planet around the Sun is an **ellipse** (like a flattened circle). Centripetal force is needed to keep the planets from moving off in a straight line into outer space. Gravity provides this centripetal force. It acts towards the centre of the Solar System, the Sun. Venus is roughly the same mass as the Earth, but travels much faster. This is possible because Venus is closer to the Sun, so the force of gravity, and therefore the centripetal force, is much larger (see p. 54).

TURNING A CORNER

FRICTION

One of the forces acting on a motorcycle as it turns a bend is the centripetal force caused by the friction between the tyres and the road. Without this friction, for example on an icy surface, a motorcycle would simply carry on in a straight line.

GYROSCOPE

ANGULAR MOMENTUM

Any spinning object, like a wheel or a top, will behave like a gyroscope. Once spinning, a gyroscope possesses **angular momentum**. This gives the gyroscope stability. The force of gravity acting on the gyroscope will not topple it. As gravity tries to tilt the axis, its axis moves at right angles to gravity's force. This causes a motion called **precession**, in which the axis traces a small circle.

Waves and oscillations

AN OSCILLATION IS ANY MOTION TO AND FRO, such as that of a pendulum. When that motion travels through matter or space, it becomes a wave. An oscillation, or vibration, occurs when a force acts that pulls a displaced object back to its **equilibrium** position, and the size of this force increases with the size of the **displacement**. A **mass** on a spring, for example, is acted upon by two forces: **gravity** and the **tension** (see pp. 28-29) in the spring. At the point of equilibrium, the **resultant** (see pp. 10-11) of these forces is zero: they cancel each other out. At all other points, the resultant force acts in a direction that restores the object to its equilibrium position. This results in the object moving to and fro, or oscillating, about that position. Vibration is very common, and results in the phenomenon of sound. In air, the vibrations that cause sound are transmitted as a wave between air molecules; many other substances transmit sound in a similar way.

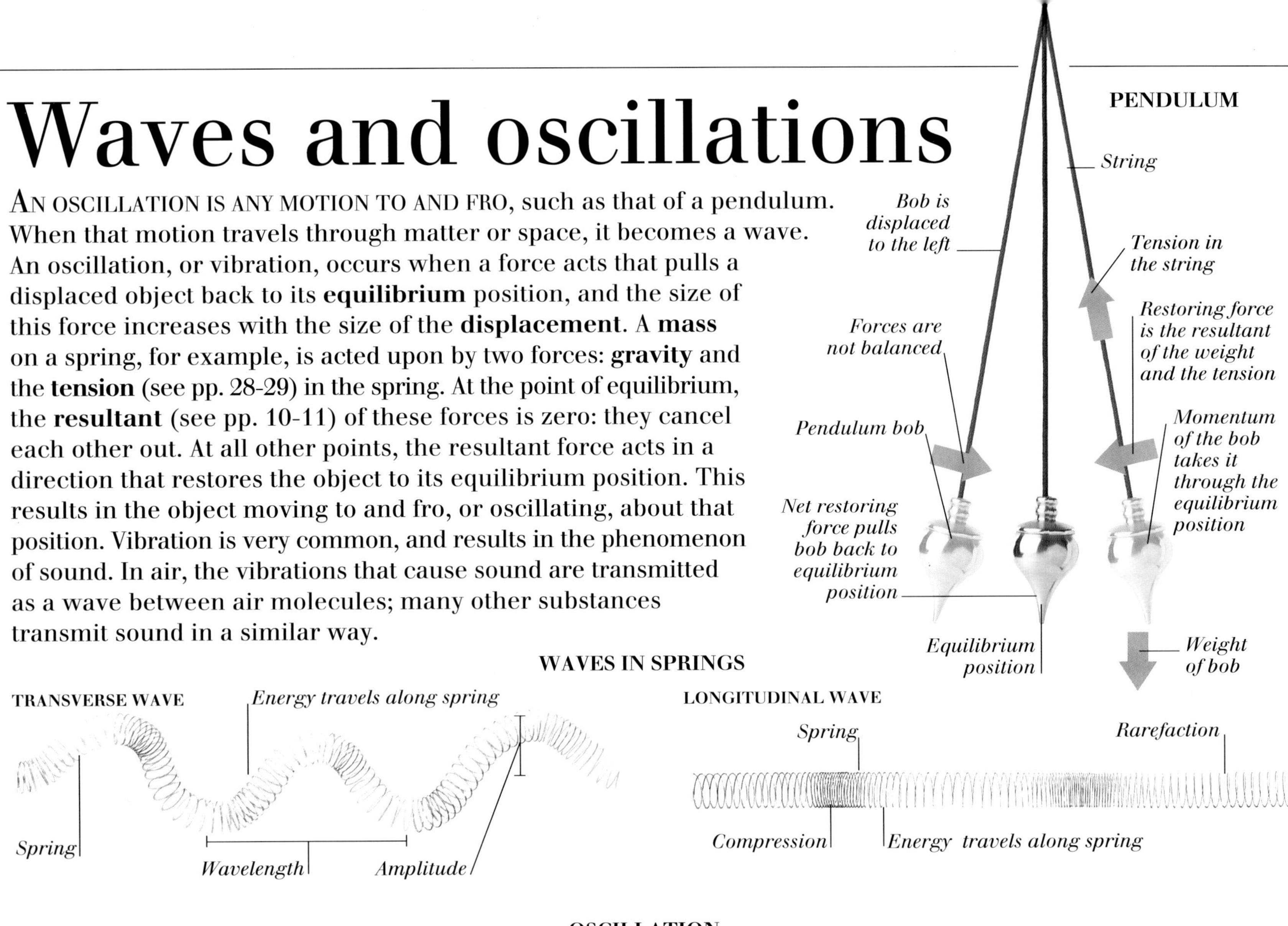

OSCILLATION

MOTION OF MASS ON SPRING

The first mass shown (below left) is in equilibrium. The two forces acting on it – its **weight** and the tension in the spring – exactly cancel each other out. The mass is given an initial downward push. Once the mass is displaced downwards (below centre), the tension in the spring exceeds the weight. The resultant upward force accelerates the mass back up towards its original position, by which time it has momentum, carrying it farther upwards. When the weight exceeds the tension in the spring (below right), the mass is pulled down again. This cycle repeats.

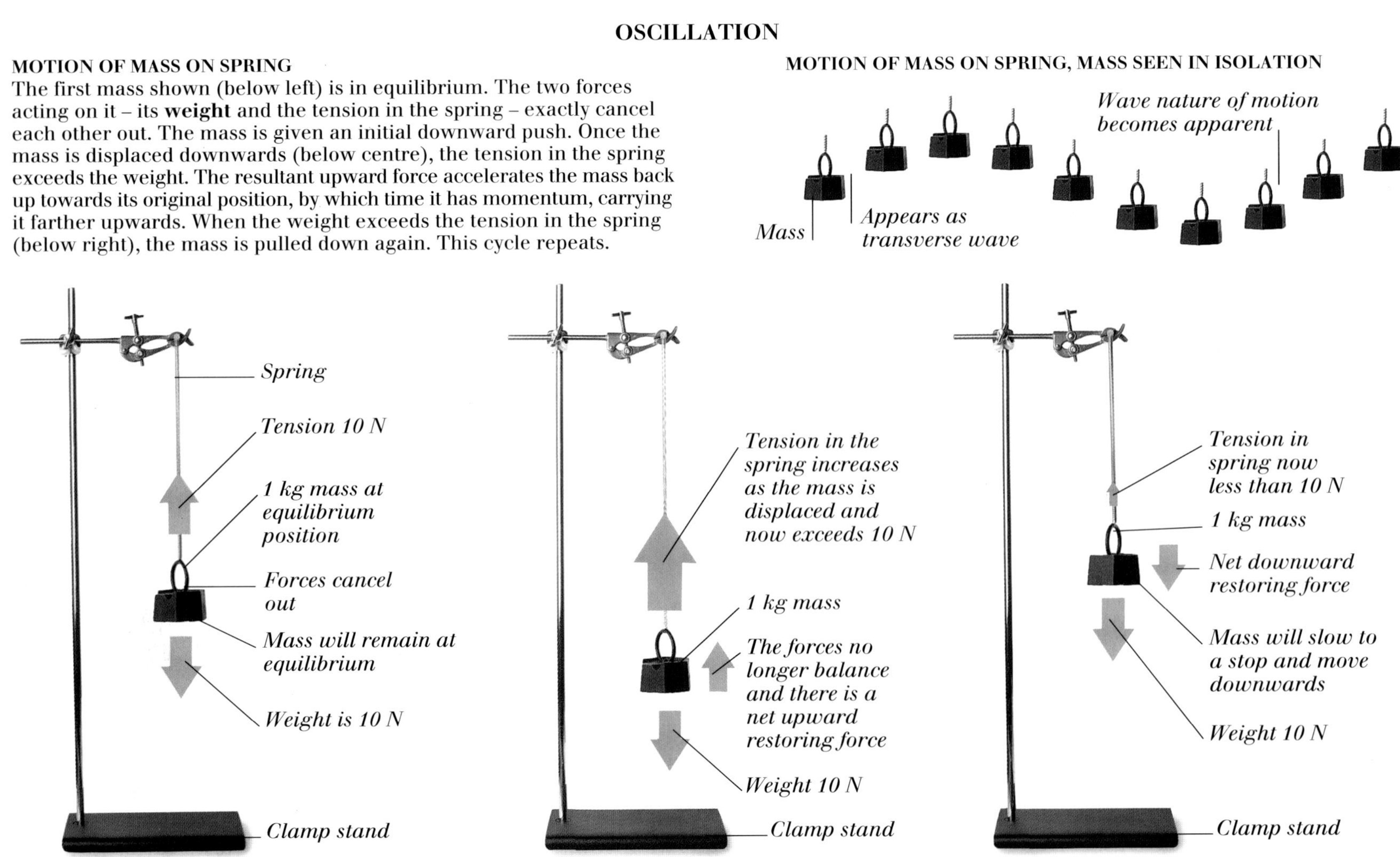

SOUND AS VIBRATION OF THE AIR

PROPAGATION OF SOUND

A vibrating object, such as the tuning fork shown here, causes variations in pressure in the surrounding air. Areas of high and low **pressure**, known as **compressions** and **rarefactions**, propagate (move) through the air as sound waves. The sound waves meet a microphone, and create electrical oscillations displayed on an oscilloscope.

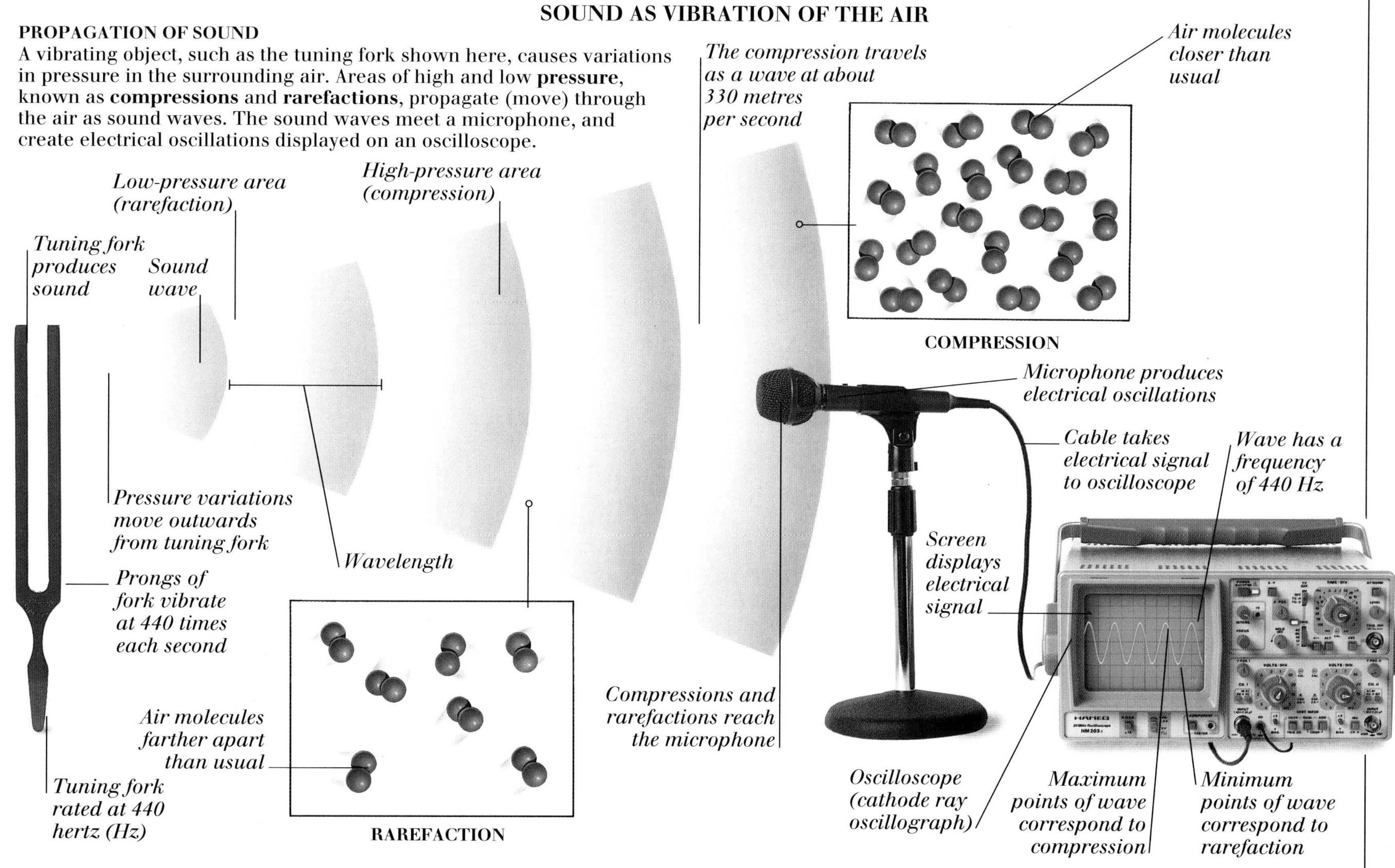

NOTES PRODUCED BY COLUMNS OF AIR

FREQUENCY AND WAVELENGTH

The distance between each compression of a sound wave is called its **wavelength**. Sound waves with a short wavelength have a high **frequency** and sound high-pitched. The frequency of a note is the number of vibrations each second, and is measured in hertz (Hz). The columns of air in these jars produce different notes when air is blown over them.

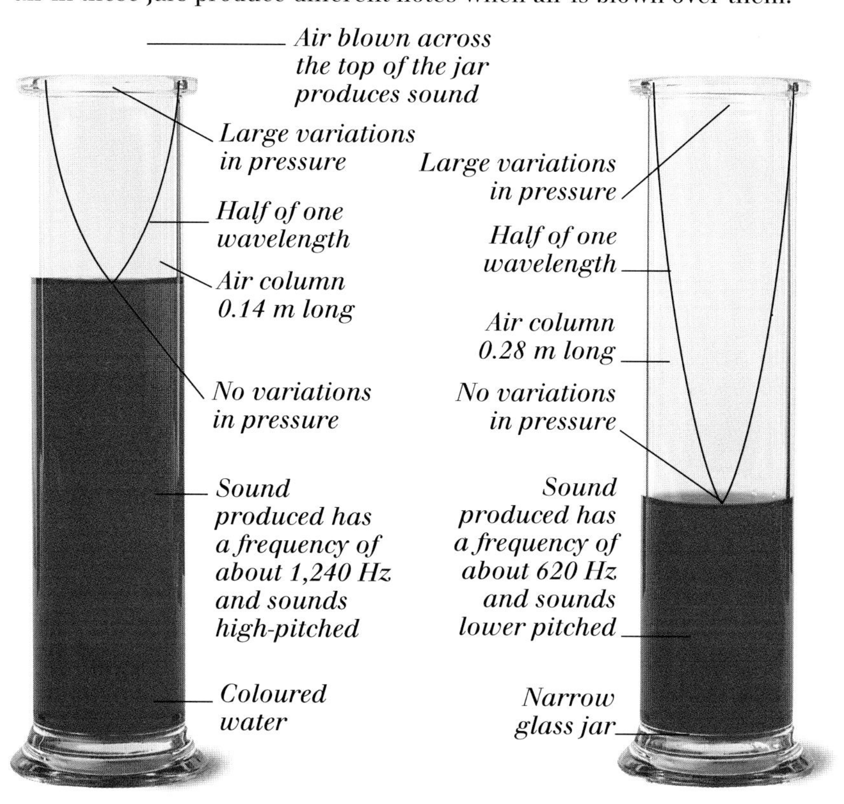

LOUDSPEAKER

A changing electrical signal is fed to the voice coil of a loudspeaker, which lies within the **magnetic field** of a **permanent magnet**. The signal in the coil causes it to behave like an **electromagnet** (see pp. 34-35), making it push against the field of the permanent magnet. The speaker cone is then pushed in and out by the coil in time with the signal.

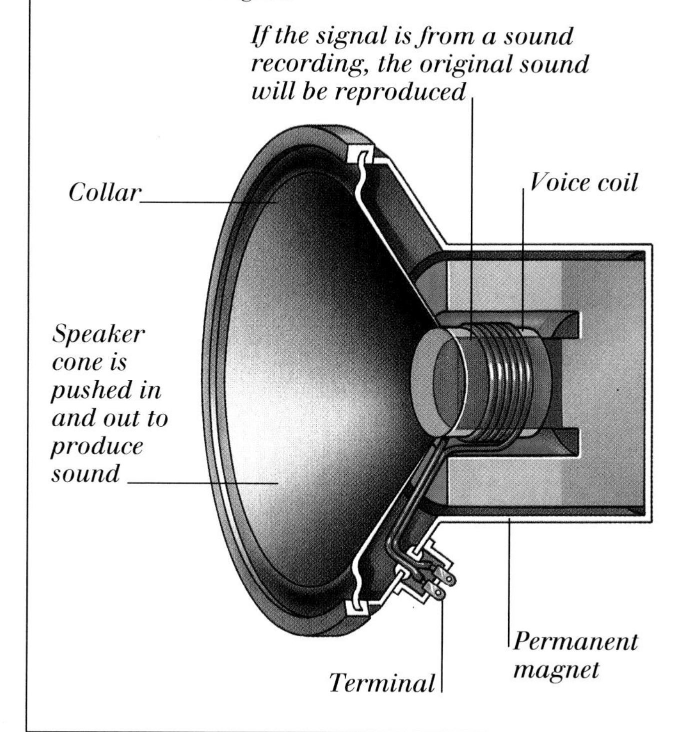

Heat and temperature

HEAT IS A FORM OF ENERGY (see pp. 6-7). This energy is the **kinetic energy** of the **atoms** and **molecules** that make up all matter. The temperature of a substance is related to the average kinetic energy of its particles. Units of temperature include the degree **Celsius**(°C), the degree **Fahrenheit** (°F), and the degree **Kelvin** (K). Some examples of equivalent values are shown below. At **absolute zero** (zero K), particles of matter do not vibrate, but at all other temperatures, particles have some motion. The **state** of a **substance** is determined by its temperature and most substances can exist as a solid (see pp. 24-25), a liquid (see pp. 26-27), or a gas (see pp. 28-29). If two substances at different temperatures make contact, their particles will share their energy. This results in a heat transfer by conduction, until the temperatures are equal. This process can melt a solid, in which case the heat transferred is called **latent heat**. Heat can also be transferred by **radiation**, in which heat energy becomes electromagnetic radiation (see pp. 38-39), and does not need a material medium to transfer heat.

RANGE OF TEMPERATURES

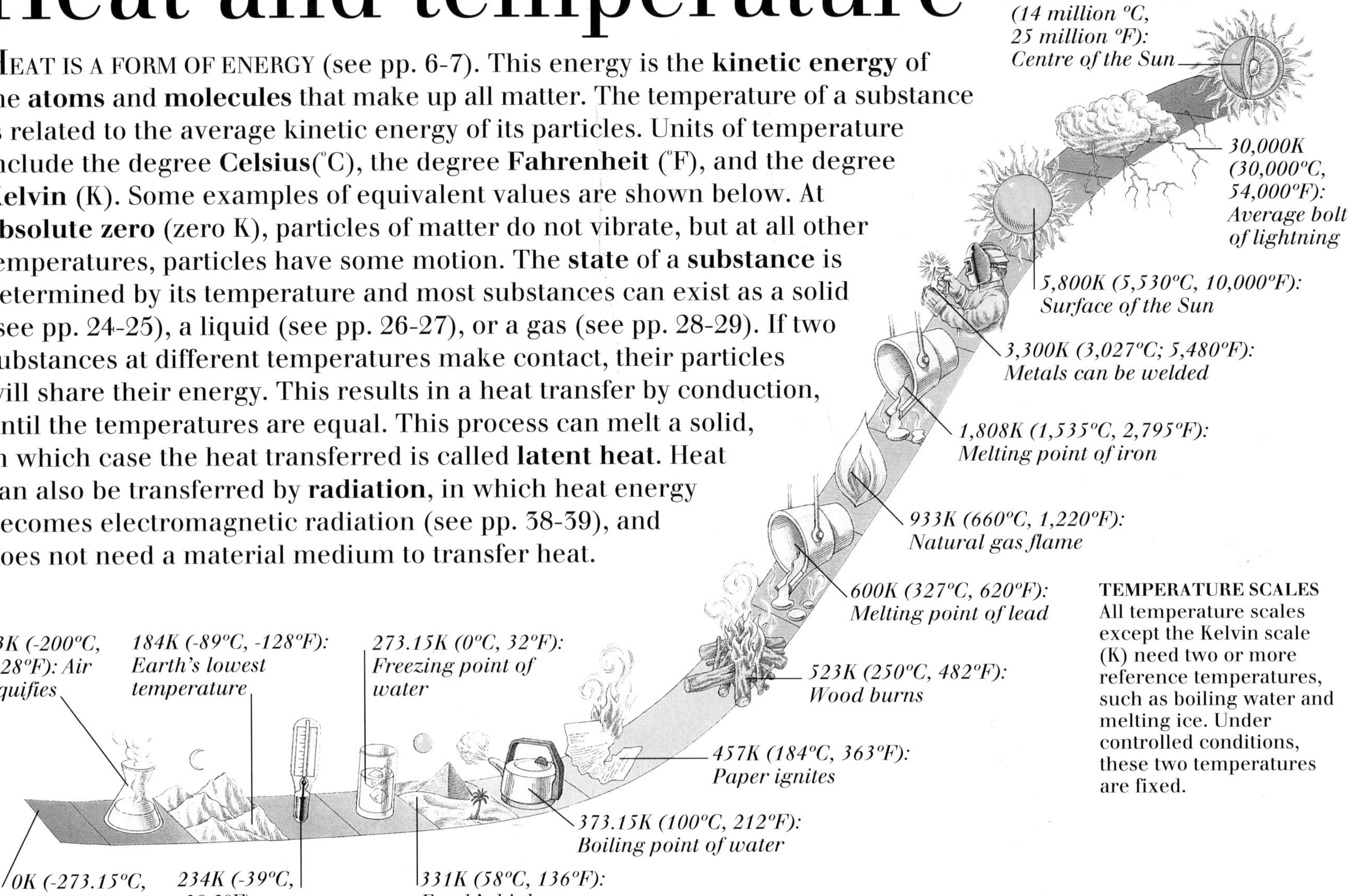

About 14 million K (14 million °C, 25 million °F): Centre of the Sun

30,000K (30,000°C, 54,000°F): Average bolt of lightning

5,800K (5,530°C, 10,000°F): Surface of the Sun

3,300K (3,027°C; 5,480°F): Metals can be welded

1,808K (1,535°C, 2,795°F): Melting point of iron

933K (660°C, 1,220°F): Natural gas flame

600K (327°C, 620°F): Melting point of lead

523K (250°C, 482°F): Wood burns

457K (184°C, 363°F): Paper ignites

373.15K (100°C, 212°F): Boiling point of water

331K (58°C, 136°F): Earth's highest temperature

273.15K (0°C, 32°F): Freezing point of water

234K (-39°C, -38.2°F): Freezing point of mercury

184K (-89°C, -128°F): Earth's lowest temperature

73K (-200°C, -328°F): Air liquifies

0K (-273.15°C, -459.67°F): Absolute zero

TEMPERATURE SCALES
All temperature scales except the Kelvin scale (K) need two or more reference temperatures, such as boiling water and melting ice. Under controlled conditions, these two temperatures are fixed.

STATES OF MATTER

GAS
Heat energy applied to a liquid allows particles to become free of each other and become a gas. However if enough energy is removed from a gas, by cooling, it condenses to a liquid.

SUPERCOOLED LIQUID
The particles of a **supercooled liquid** are in fixed positions, like those of a solid, but they are disordered and cannot be called a true solid. Supercooled liquids flow very slowly, and have no definite melting point.

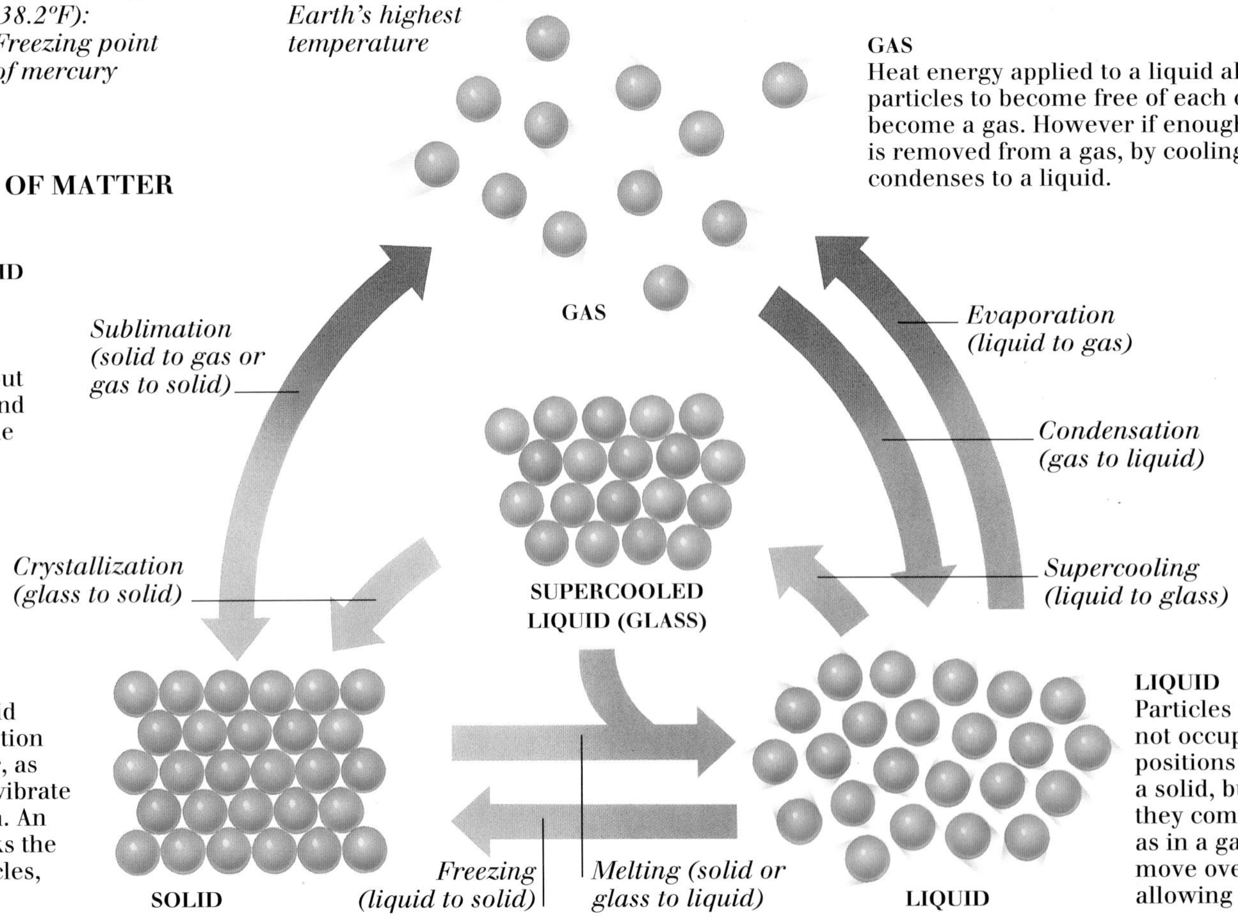

GAS

Sublimation (solid to gas or gas to solid)

Evaporation (liquid to gas)

Condensation (gas to liquid)

Supercooling (liquid to glass)

Crystallization (glass to solid)

SUPERCOOLED LIQUID (GLASS)

SOLID
The particles of a solid normally have no motion relative to each other, as they are only free to vibrate about a fixed position. An input of energy breaks the bonds between particles, and the solid melts.

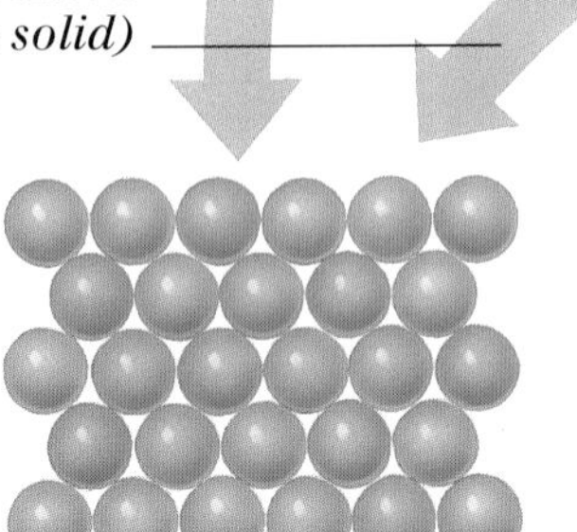

SOLID

Freezing (liquid to solid)

Melting (solid or glass to liquid)

LIQUID

LIQUID
Particles in a liquid do not occupy fixed positions like those in a solid, but neither are they completely free, as in a gas. The particles move over one another, allowing a liquid to flow.

EQUALIZATION OF TEMPERATURES

OBJECTS AT DIFFERENT TEMPERATURES

The particles of objects at different temperatures have different kinetic energies. The colours of the blocks below are an indication of their temperature.

TRANSFER OF HEAT

When two objects at different temperatures are brought into contact, a transfer of kinetic energy takes place in the form of heat. Here, the hot and cold blocks are touching.

EQUAL TEMPERATURES

Eventually, the average kinetic energies of particles in two touching objects become equal. The temperatures of the two objects are then said to be equal, as shown by the blocks below.

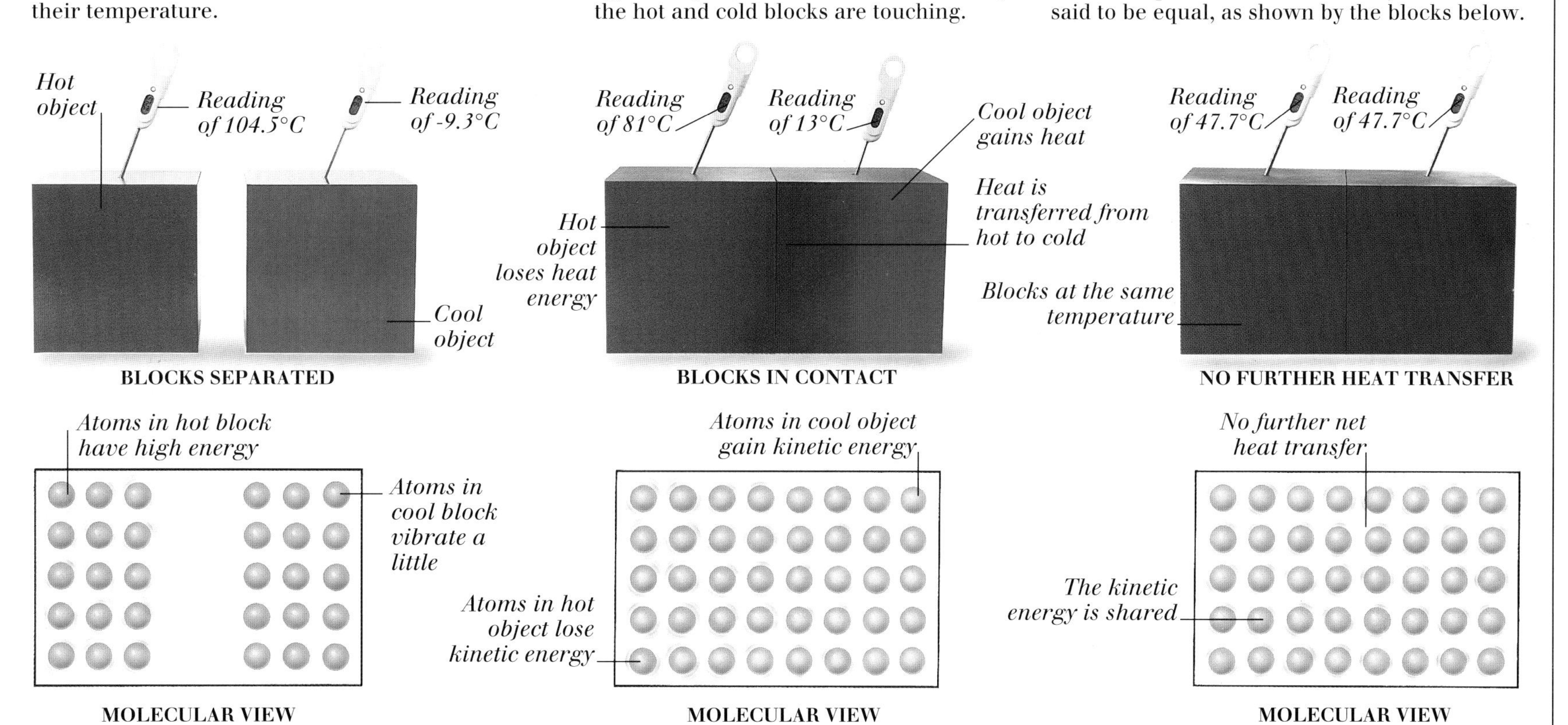

BLOCKS SEPARATED | BLOCKS IN CONTACT | NO FURTHER HEAT TRANSFER

MOLECULAR VIEW | MOLECULAR VIEW | MOLECULAR VIEW

LATENT HEAT

HEATING A SUBSTANCE

Heat transferred from a hot flame to a cooler substance can cause the substance to melt. The temperature of the substance (here, naphthalene) rises with the transfer of more energy, until it reaches the **melting point**.

Thermometer reads 80.5°C

Temperature stays the same during melting

Beaker

Gauze

Liquid naphthalene

Solid naphthalene

Bunsen burner

Hot flame

MELTING A SUBSTANCE

At the melting point, the supplied energy must break the attraction between all the particles, melting all the solid, before the temperature will rise again. This extra supplied energy is called **latent heat**.

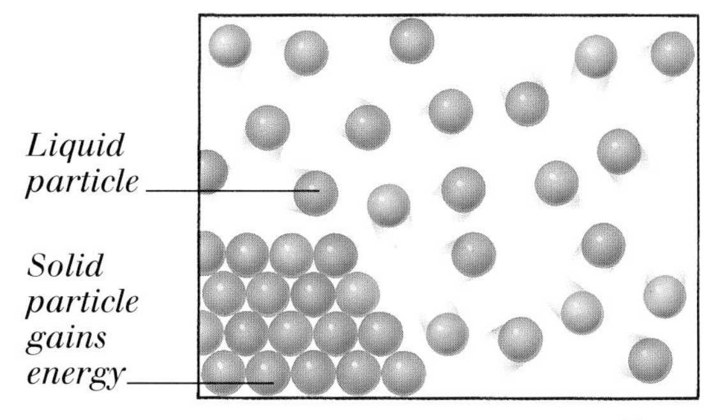

MELTING

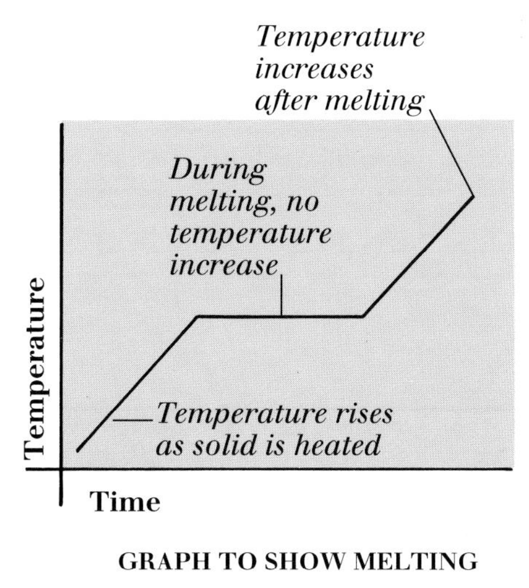

GRAPH TO SHOW MELTING

TRANSFER OF HEAT BY RADIATION

An object at room temperature produces radiation – called **infrared radiation**. A hot object, such as the lamp below, produces a lot of infrared. This radiation can heat up other objects. The hot object cools as it loses energy as radiation.

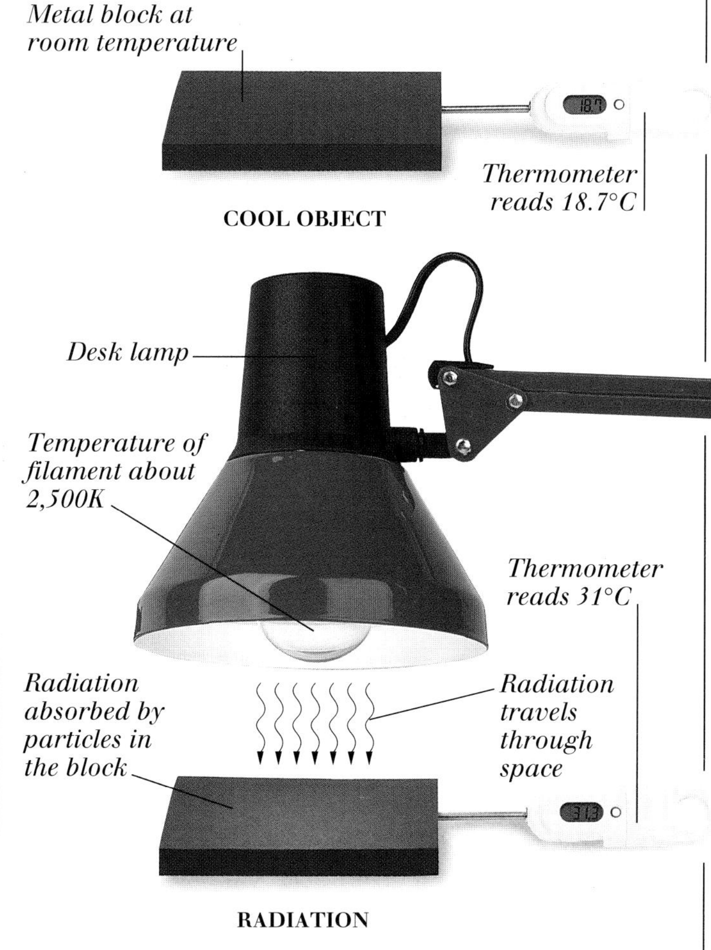

COOL OBJECT

RADIATION

Solids

THE ATOMS OF A SOLID ARE CLOSELY PACKED, giving it a greater **density** than most liquids and all gases. A solid's rigidity derives from the strong attraction between its atoms. A force pulling on a solid moves these atoms farther apart, creating an opposing force called **tension**. If a force pushes on a solid, the atoms move closer together, creating **compression**. Temperature (see pp. 22-23) can also affect the nature of a solid. When the temperature of a solid increases, its particles gain **kinetic energy** and vibrate more vigorously, resulting in **thermal expansion**. Most solids are **crystals**, in which atoms are arranged in one of seven regular, repeating patterns (see below). **Amorphous solids**, such as glass, are not composed of crystals and can be moulded into any shape. When the atoms of a solid move apart, the length of the solid increases. The extent of this increase depends on the applied force, and on the thickness of the material, and is known as **elasticity**.

STEEL RAILS

The expansion of a solid with an increase in temperature (see below) would cause rails to buckle badly in hot weather. To prevent this, rails are made in sections. The gap between the two sections allows each section to expand without buckling.

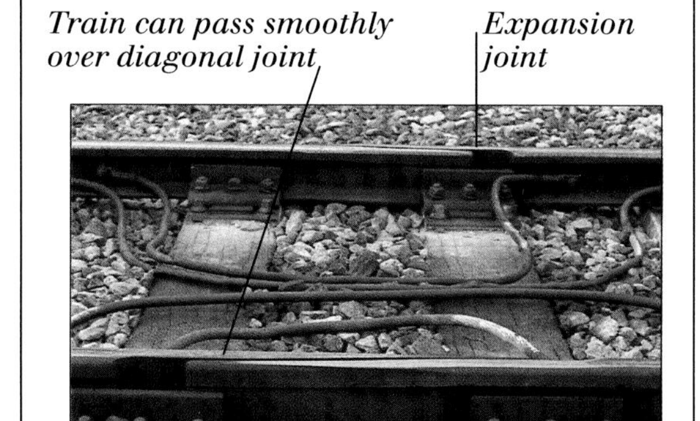

THERMAL EXPANSION

EXPERIMENT TO SHOW THERMAL EXPANSION
When a substance is heated, its atoms gain kinetic energy. In a solid, this results in the atoms vibrating more vigorously about their fixed positions. As a result, solids expand when heated. Below, a thin steel rod is heated by a gas flame, and the resulting expansion is measured using a **micrometer**.

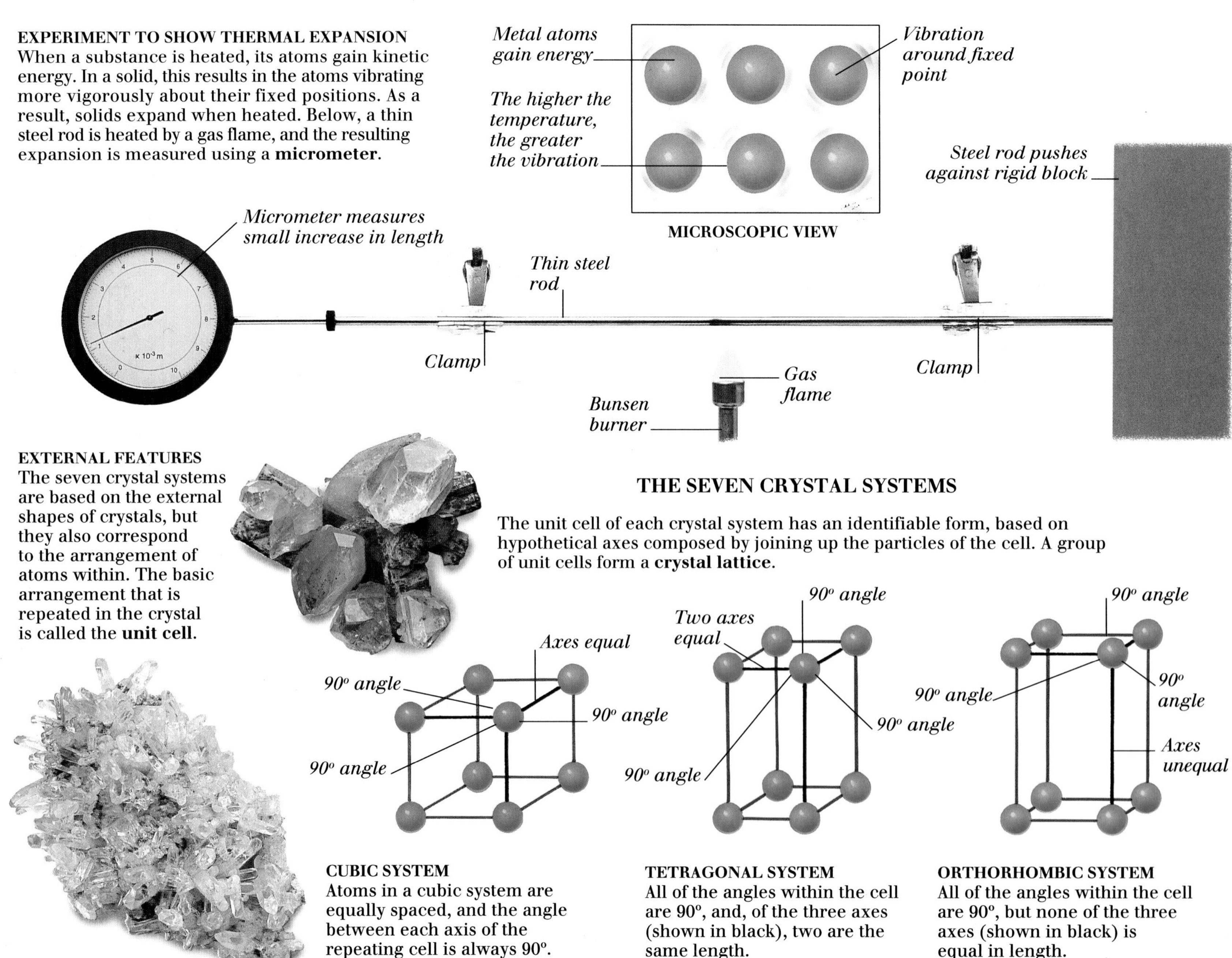

EXTERNAL FEATURES
The seven crystal systems are based on the external shapes of crystals, but they also correspond to the arrangement of atoms within. The basic arrangement that is repeated in the crystal is called the **unit cell**.

THE SEVEN CRYSTAL SYSTEMS

The unit cell of each crystal system has an identifiable form, based on hypothetical axes composed by joining up the particles of the cell. A group of unit cells form a **crystal lattice**.

CUBIC SYSTEM
Atoms in a cubic system are equally spaced, and the angle between each axis of the repeating cell is always 90°.

TETRAGONAL SYSTEM
All of the angles within the cell are 90°, and, of the three axes (shown in black), two are the same length.

ORTHORHOMBIC SYSTEM
All of the angles within the cell are 90°, but none of the three axes (shown in black) is equal in length.

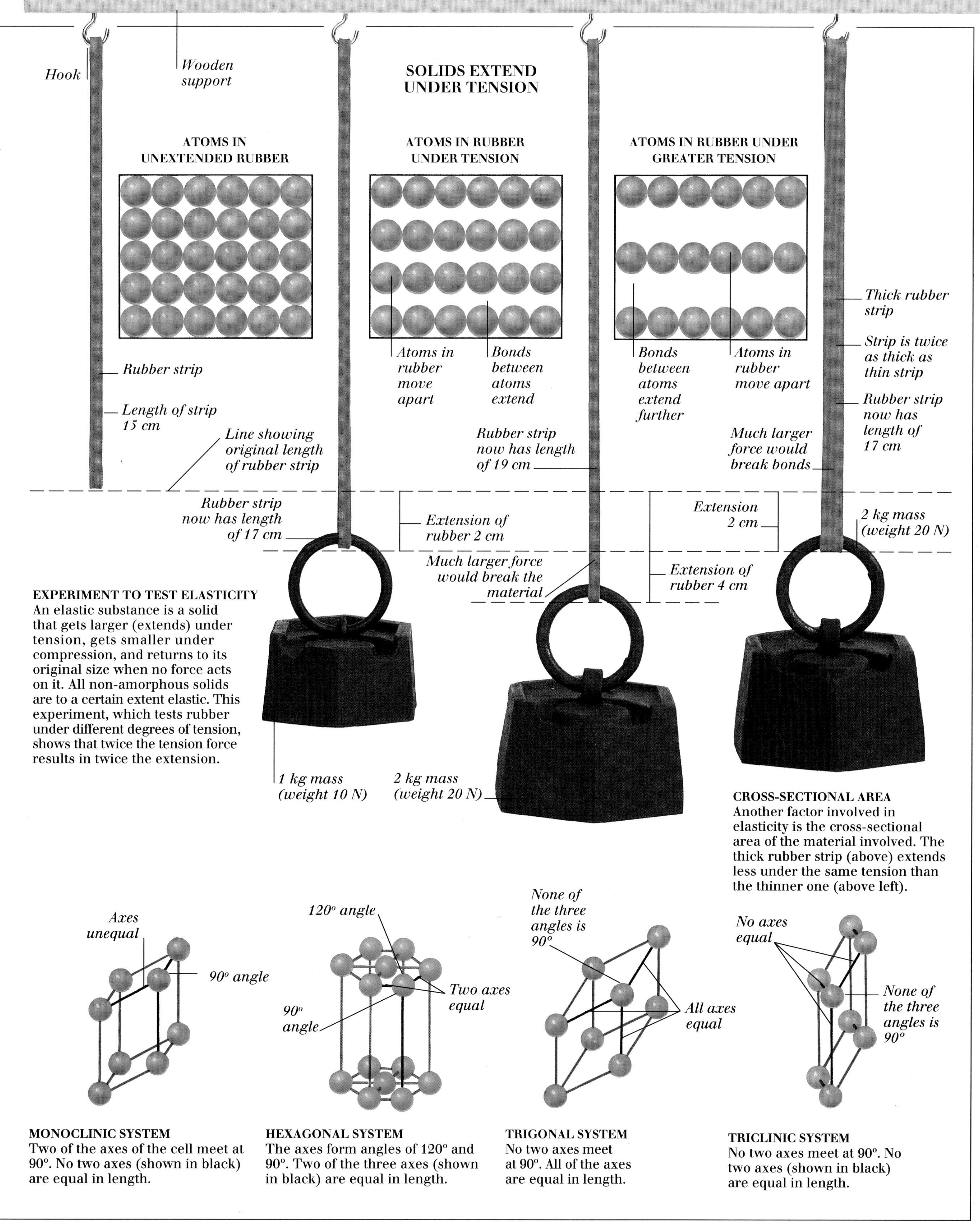

EXPERIMENT TO TEST ELASTICITY
An elastic substance is a solid that gets larger (extends) under tension, gets smaller under compression, and returns to its original size when no force acts on it. All non-amorphous solids are to a certain extent elastic. This experiment, which tests rubber under different degrees of tension, shows that twice the tension force results in twice the extension.

CROSS-SECTIONAL AREA
Another factor involved in elasticity is the cross-sectional area of the material involved. The thick rubber strip (above) extends less under the same tension than the thinner one (above left).

MONOCLINIC SYSTEM
Two of the axes of the cell meet at 90°. No two axes (shown in black) are equal in length.

HEXAGONAL SYSTEM
The axes form angles of 120° and 90°. Two of the three axes (shown in black) are equal in length.

TRIGONAL SYSTEM
No two axes meet at 90°. All of the axes are equal in length.

TRICLINIC SYSTEM
No two axes meet at 90°. No two axes (shown in black) are equal in length.

Liquids

UNLIKE SOLIDS, LIQUIDS CAN FLOW. Their particles move almost independently of each other but are not as free as the particles of a gas. Forces of attraction called **cohesive** forces act between the particles of a liquid. These forces create **surface tension**, which pulls liquid drops into a spherical shape. If the surface tension of water is reduced, by dissolving soap in it, then pockets of air can stretch the surface into a thin film, forming a bubble. Forces of attraction between liquid particles and adjoining matter are called **adhesive** forces. The balance between cohesive and adhesive forces causes **capillary action**, and the formation of a **meniscus** curve at the boundary between a liquid and its container. Liquids exert pressure on any object immersed in them; the pressure acts in all directions and increases with depth, creating **upthrust** on an immersed object. If the upthrust is large enough, the object will float.

LIQUID DROPS AND BUBBLES

COHESIVE FORCES

No **resultant** force acts on any particle within the liquid, because cohesive forces pull it in every direction. But at the surface, the resultant force on each particle pulls it inwards. This causes surface tension, which pulls drops and bubbles into spheres. A water drop on a surface will be flattened slightly by gravity.

Surface tension

SPHERICAL SOAP BUBBLE

Curved surface of drop

WATER DROP ON A SURFACE

Cohesive forces act in all directions

Curved surface of drop

Surface particle

Particle within liquid

SURFACE TENSION

LIQUIDS IN TUBES

CAPILLARY ACTION

Water adheres to glass. This adhesion can lift water up into a glass tube; an effect known as capillary action.

Narrow 0.5 mm capillary tube

4 mm diameter glass tube

5 mm diameter glass tube

Water level

Water is lifted higher in a narrow tube than in a wide one because the narrow column of water weighs less

Water level

Water level

Shallow glass dish

Body of liquid

MENISCUS

Where a liquid meets a solid surface, a curve called a meniscus forms. The shape of the meniscus depends on the balance between cohesive and adhesive forces.

DOWNWARD MENISCUS

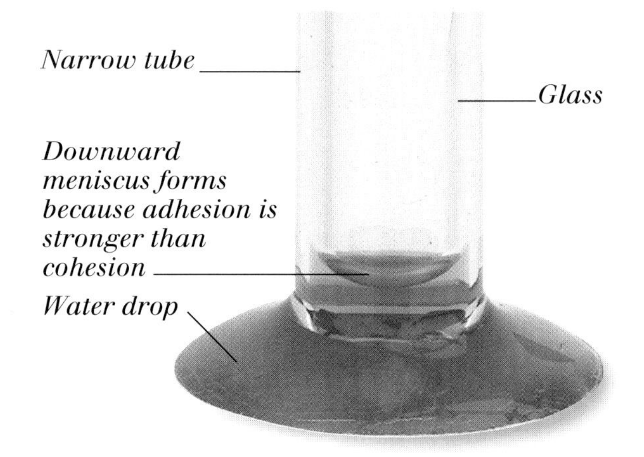

Wall of glass tube

MOLECULAR VIEW

Capillary action is caused by adhesive and cohesive forces between particles of glass and water. Here, water molecules adhere to glass and the adhesive force lifts the edge of the water up the glass. The cohesive forces between water molecules means that this lifted edge also raises water molecules lying farther out from the edge of the glass.

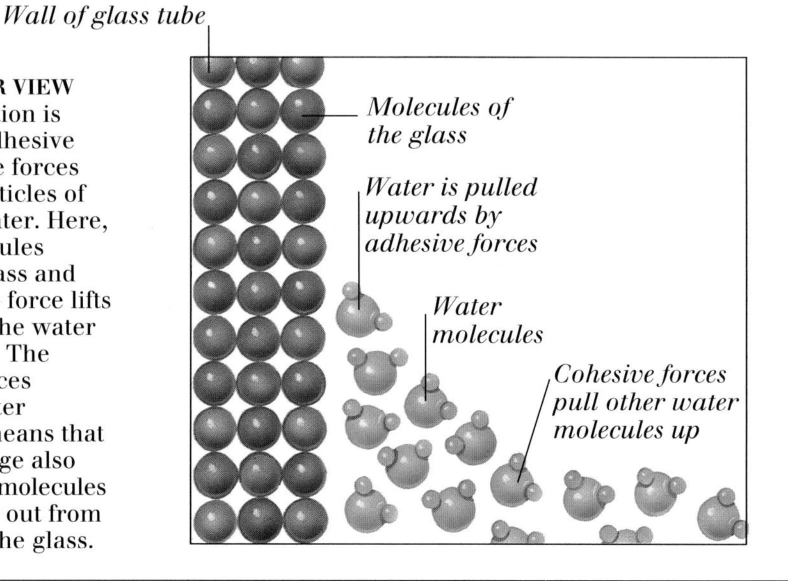

UPWARD MENISCUS

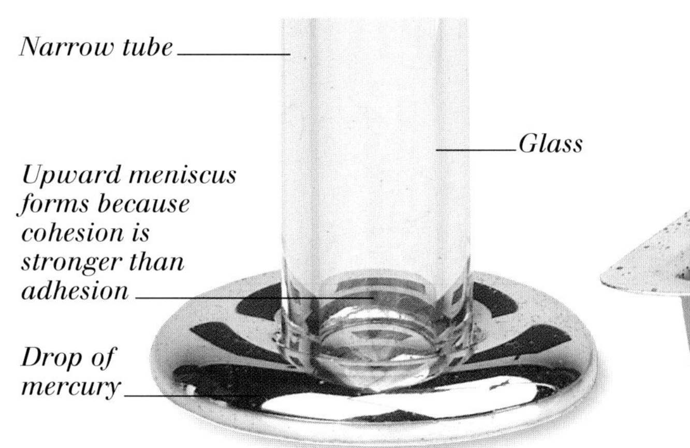

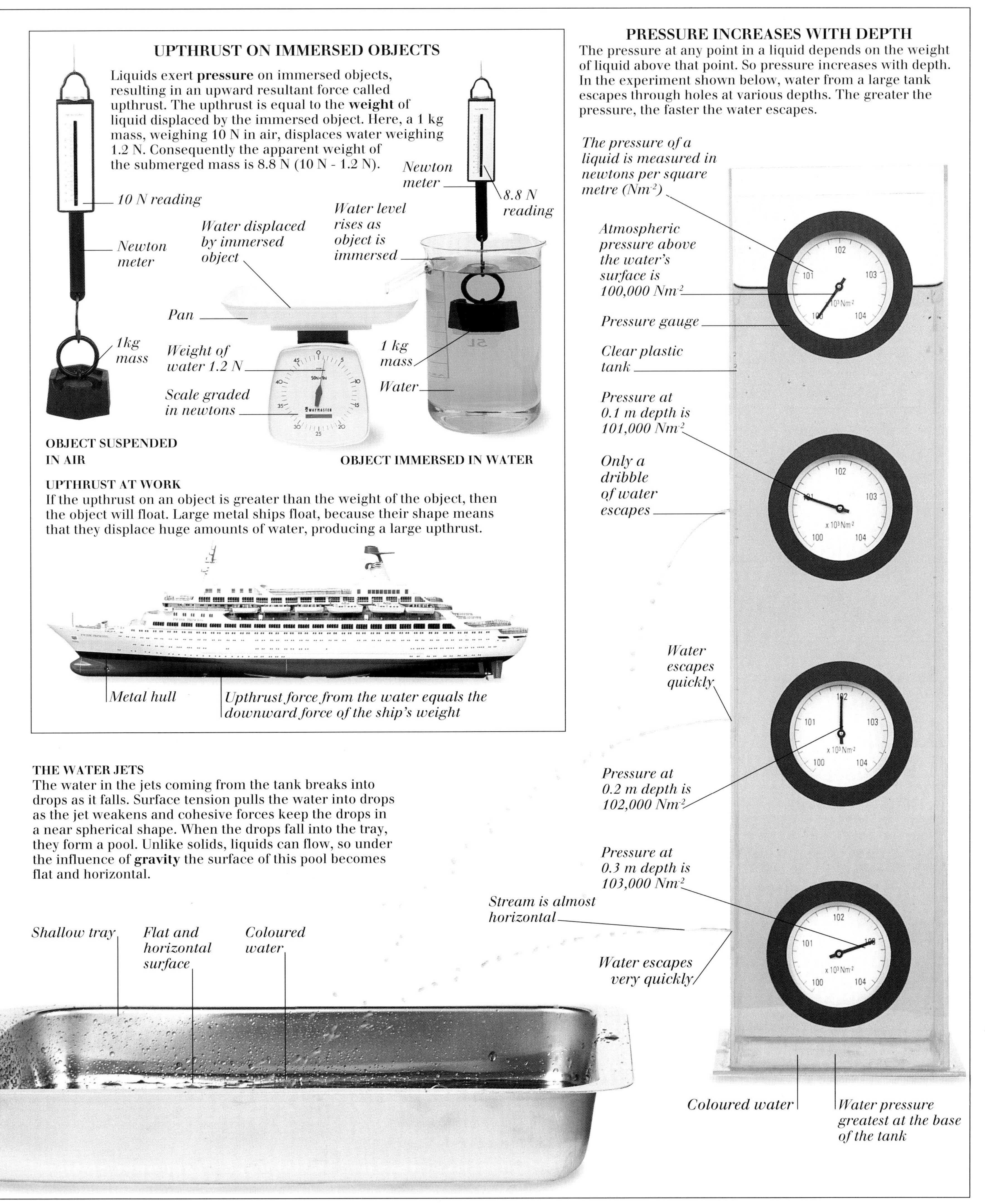

UPTHRUST ON IMMERSED OBJECTS

Liquids exert **pressure** on immersed objects, resulting in an upward resultant force called upthrust. The upthrust is equal to the **weight** of liquid displaced by the immersed object. Here, a 1 kg mass, weighing 10 N in air, displaces water weighing 1.2 N. Consequently the apparent weight of the submerged mass is 8.8 N (10 N - 1.2 N).

UPTHRUST AT WORK

If the upthrust on an object is greater than the weight of the object, then the object will float. Large metal ships float, because their shape means that they displace huge amounts of water, producing a large upthrust.

THE WATER JETS

The water in the jets coming from the tank breaks into drops as it falls. Surface tension pulls the water into drops as the jet weakens and cohesive forces keep the drops in a near spherical shape. When the drops fall into the tray, they form a pool. Unlike solids, liquids can flow, so under the influence of **gravity** the surface of this pool becomes flat and horizontal.

PRESSURE INCREASES WITH DEPTH

The pressure at any point in a liquid depends on the weight of liquid above that point. So pressure increases with depth. In the experiment shown below, water from a large tank escapes through holes at various depths. The greater the pressure, the faster the water escapes.

Gases

A GAS COMPRISES INDEPENDENT PARTICLES – **atoms** or **molecules** – in random motion. This means that a gas will fill any container into which it is placed. If two different gases are allowed to meet, the particles of the gases will mix together. This process is known as **diffusion**. Imagine a fixed mass of gas – that is, a fixed number of gas particles. It will occupy a particular amount of space, or **volume**, often confined by a container. The particles of the gas will be in constant, random motion. The higher the **temperature** of the gas (see pp. 22-23), the faster the particles move. The bombardment of particles against the sides of the container produces **pressure** (see pp. 10-11). Three simple laws describe the predictable behaviour of gases. They are Boyle's Law, the Pressure Law, and Charles' Law. Each of the gas laws describes a relationship between the pressure, volume, and temperature of a gas.

Clear perspex shield

Closed glass tube

Thick tube wall withstands pressure

Volume of trapped air

Column of oil

After each pressure change, apparatus is allowed to revert to room temperature

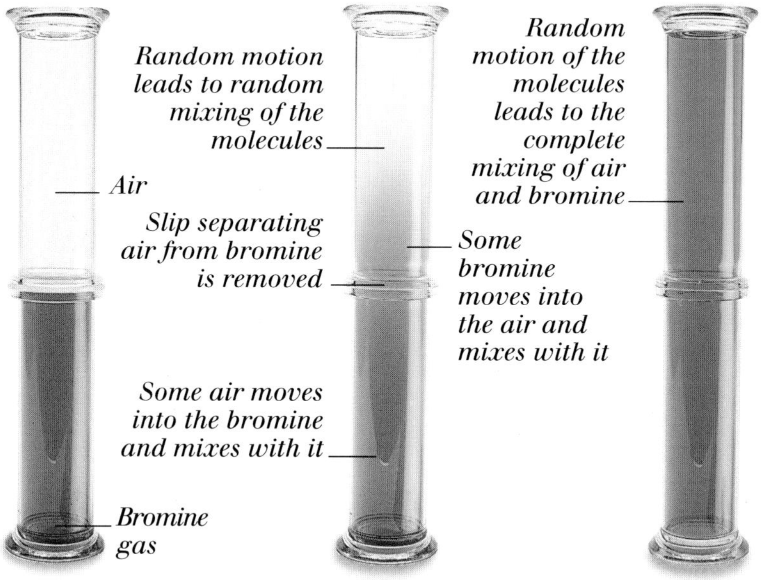

BOYLE'S LAW

The volume of a **mass** of gas at a fixed temperature will change in relation to the pressure. If the pressure on a gas increases, its volume will decrease. The apparatus on the left is used to illustrate Boyle's Law. A foot pump is used to push a column of oil up a sealed tube, reducing the volume occupied by the gas in the top part of the tube.

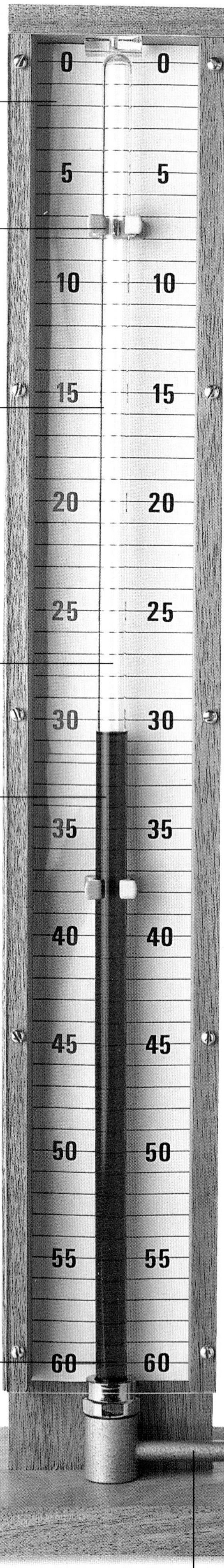

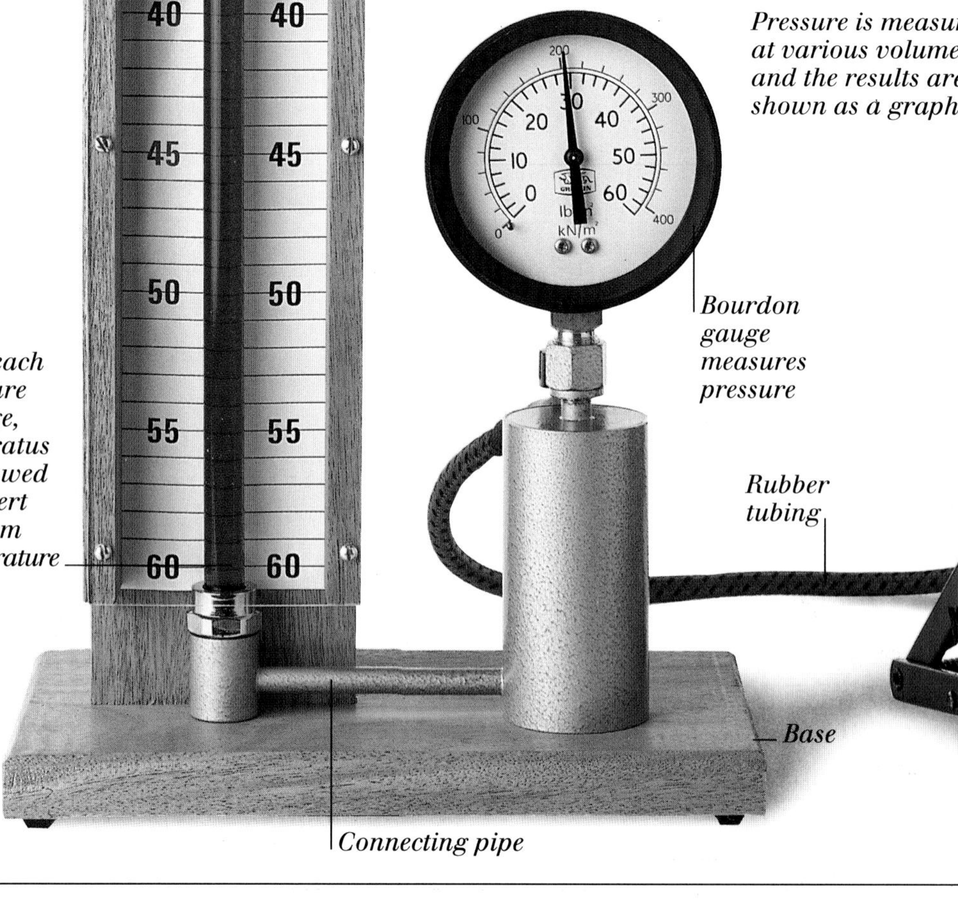

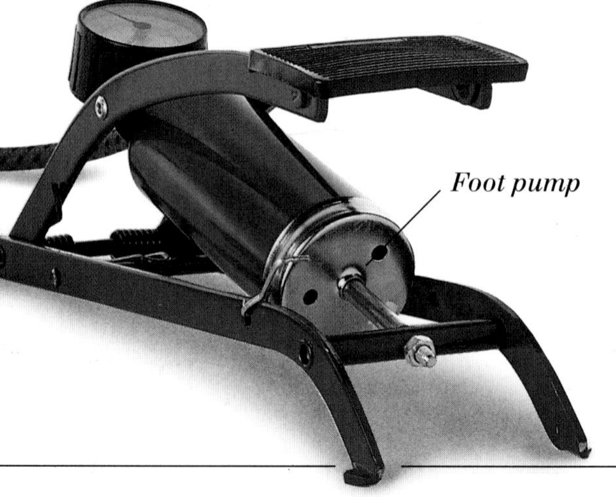

DIFFUSION

The random movement of gas particles ensures that any two gases sharing the same container will totally mix. This is diffusion. In the experiment below, the lower gas jar contains bromine, the top one air.

Air

Bromine gas

Random motion leads to random mixing of the molecules

Slip separating air from bromine is removed

Some air moves into the bromine and mixes with it

Some bromine moves into the air and mixes with it

Random motion of the molecules leads to the complete mixing of air and bromine

Pressure is measured at various volumes and the results are shown as a graph

GRAPH OF PRESSURE AND VOLUME READINGS

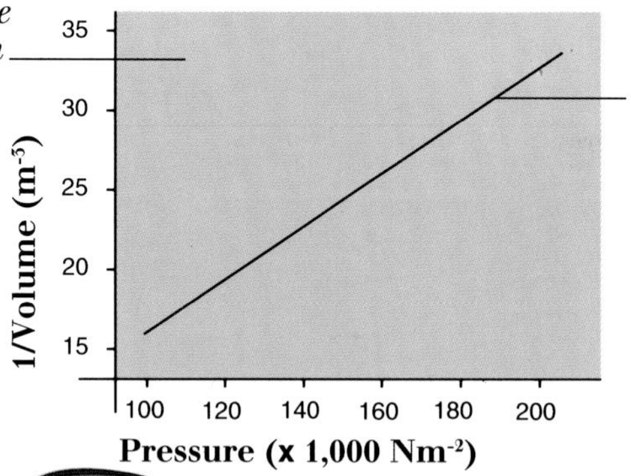

1/Volume (m^{-3})

Pressure (x 1,000 Nm^{-2})

Doubling the pressure halves the volume

Bourdon gauge measures pressure

Rubber tubing

Foot pump

Base

Connecting pipe

PRESSURE LAW

The pressure exerted by a gas at constant volume increases as the temperature of the gas rises. The apparatus shown is used to verify the Pressure Law. A mass of gas is heated in a water bath, and the pressure of the gas measured. When plotted as points on a graph the results lie on a straight line.

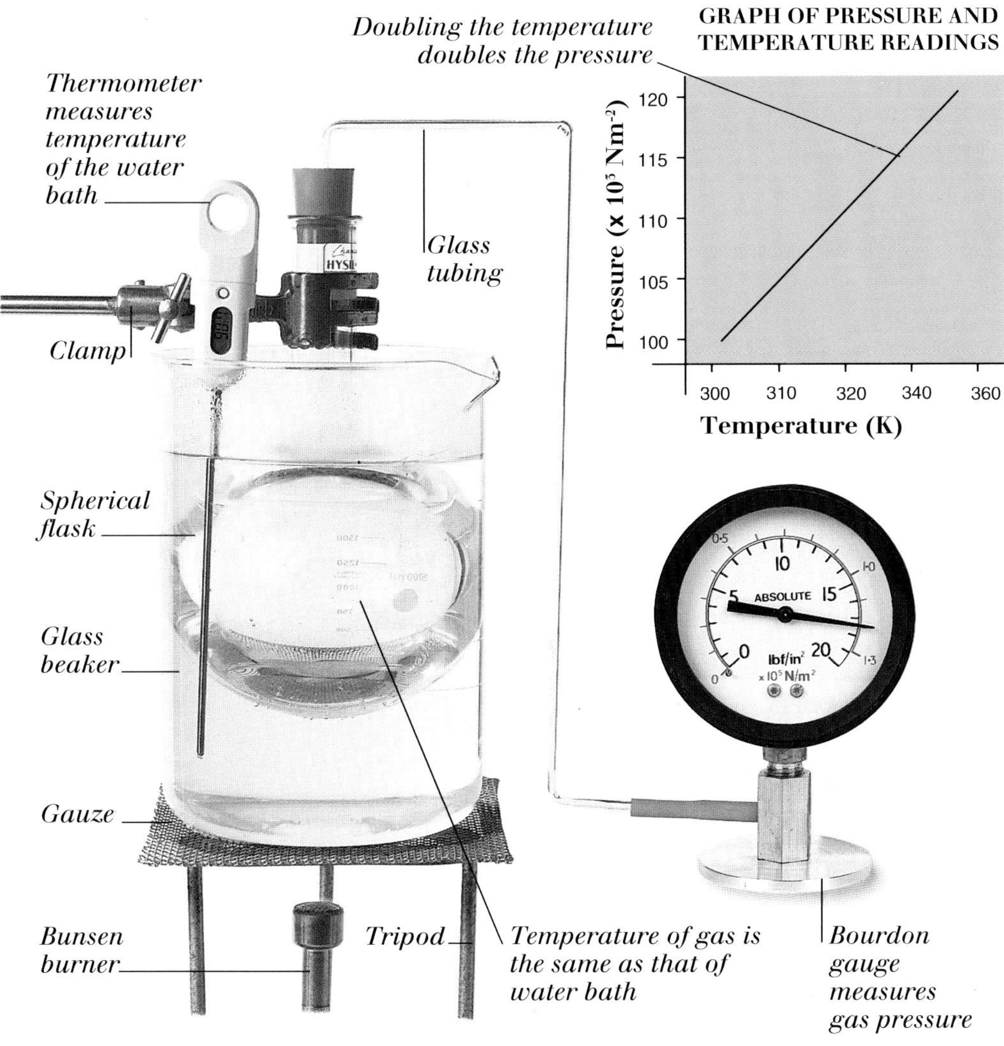

CHARLES' LAW

The volume of a mass of gas at a fixed pressure depends on its temperature. The higher the temperature, the greater the volume. The apparatus shown is used to illustrate Charles' Law. The volume of a gas sample in the glass bulb is noted at various temperatures. A graph shows the results.

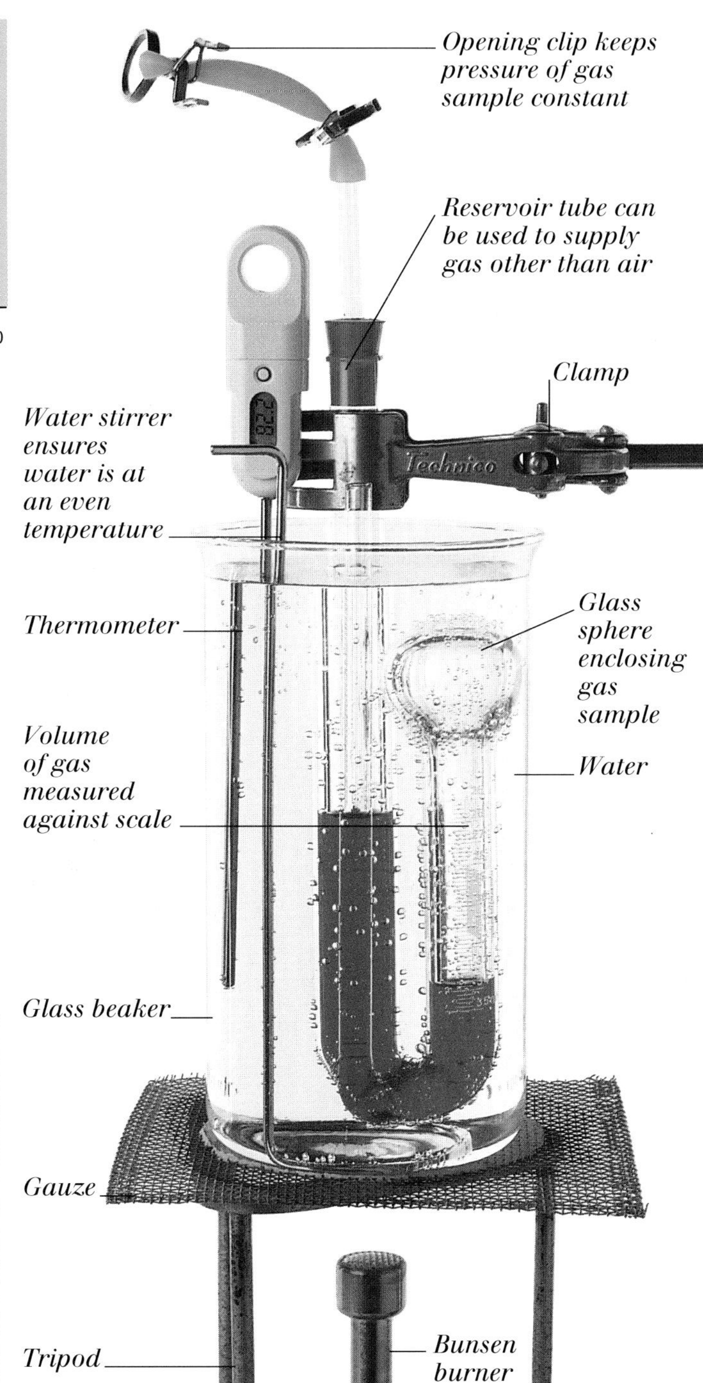

HOT-AIR BALLOON – CHARLES' LAW IN ACTION

The air in the envelope of a hot-air balloon is heated by a gas burner. As its temperature rises, the gas expands in accordance with Charles' Law. The envelope is open at the bottom, so some hot air escapes. Because air has mass (and therefore **weight**), the balloon weighs less once some air has escaped, although its volume is still large. The pressure of the air outside the envelope produces an **upthrust,** which, if enough air has been lost from the envelope, will be great enough to lift the balloon.

Envelope

Hot air escapes

Gas burner

Basket

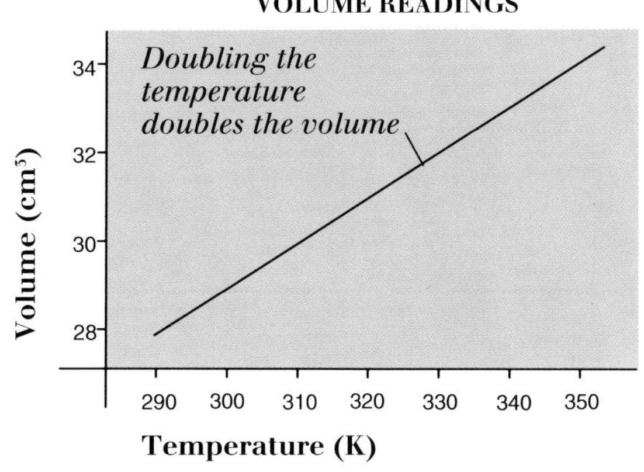

Electricity and magnetism

ALL ELECTRICAL EFFECTS ARE CAUSED by **electric charges.** There are two types of electric charge, positive and negative. These charges exert **electrostatic** forces on each other. An **electric field** is the region in which these forces have effect. In atoms, **protons** (see pp. 48-49) carry positive charge, while **electrons** carry negative charge. **Atoms** are normally neutral, having equal numbers of each charge, but an atom can gain or lose electrons, for example by being rubbed. It then becomes a charged atom, or **ion**. Ions can be produced continuously by a Van de Graaff generator. Ions in a charged object may cause another nearby object to become charged. This process is called **induction**. Electricity has many similarities with magnetism (see pp. 34-35). For example, the lines of the electric field between charges (see right) take the same form as lines of magnetic force (see opposite), so magnetic fields are equivalent to electric fields. Iron consists of small magnetized regions called **domains**. If the magnetic directions of the domains in a piece of iron line up, the iron becomes magnetized.

ELECTRIC FIELDS AND FORCES

Charges of the same type repel, while charges of a different type attract. One way to think of an electric field is as a set of lines of force, as illustrated below.

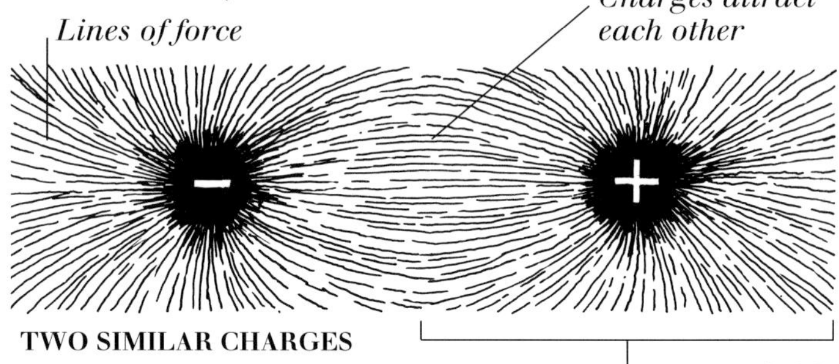

TWO SIMILAR CHARGES

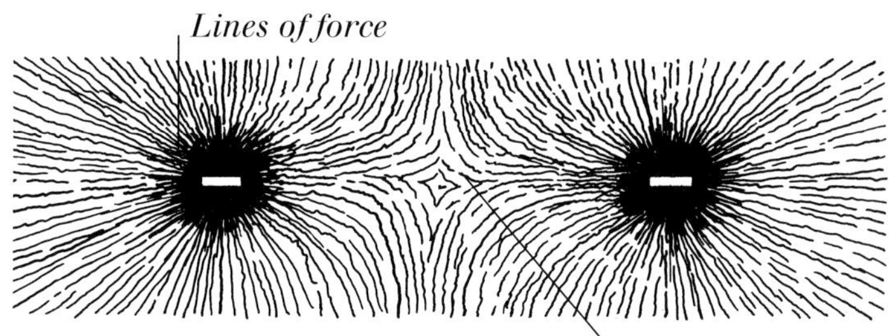

TWO DIFFERENT CHARGES

STATIC ELECTRICITY

GOLD LEAF ELECTROSCOPE

A polythene rod can gain extra electrons when it is rubbed. Touching the charged rod to the top of an electroscope causes electrons to move into the electroscope. The electrons in the central strip and in the gold leaf repel each other, and the leaf lifts.

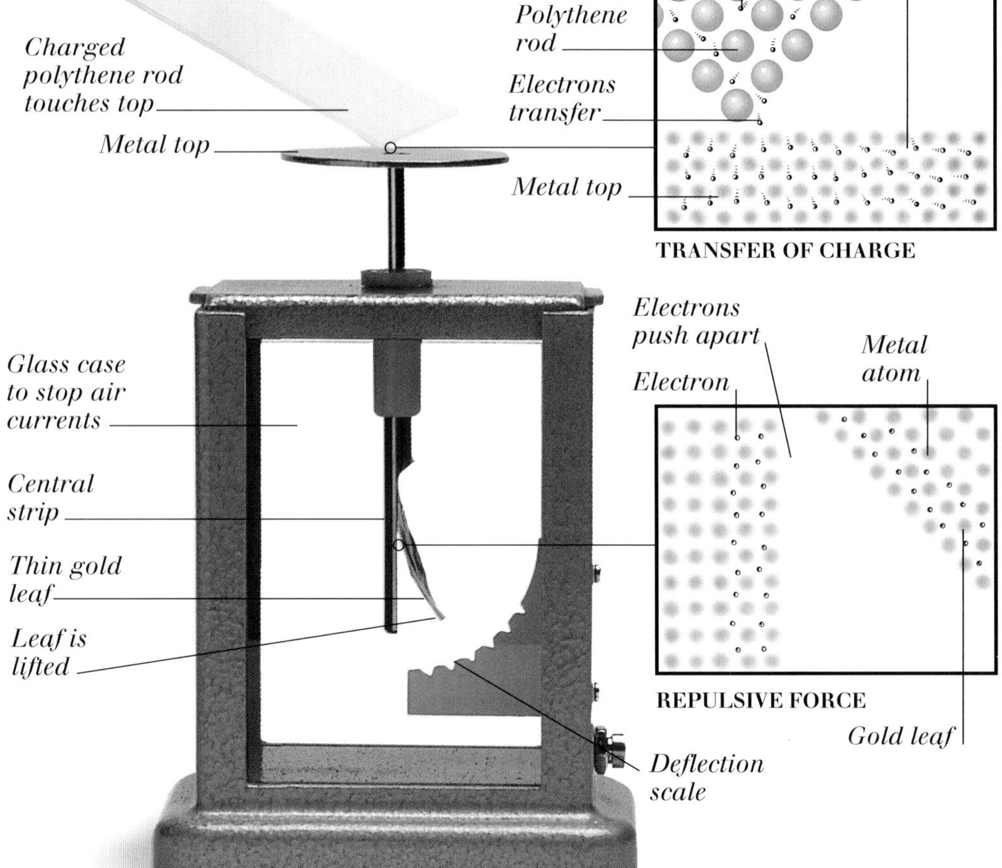

INDUCTION

When a charged object is brought near to other materials, such as paper, electrostatic forces cause a displacement of charge within that material. This is called induction. Negative charges in the paper are displaced so the edge of the paper nearest the rod becomes positively charged and clings to the negatively charged rod.

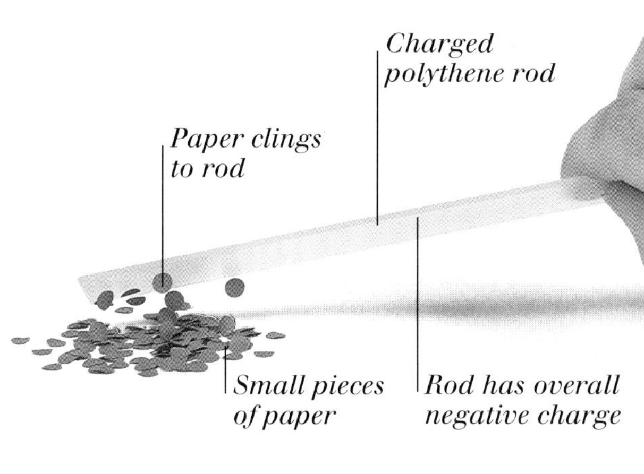

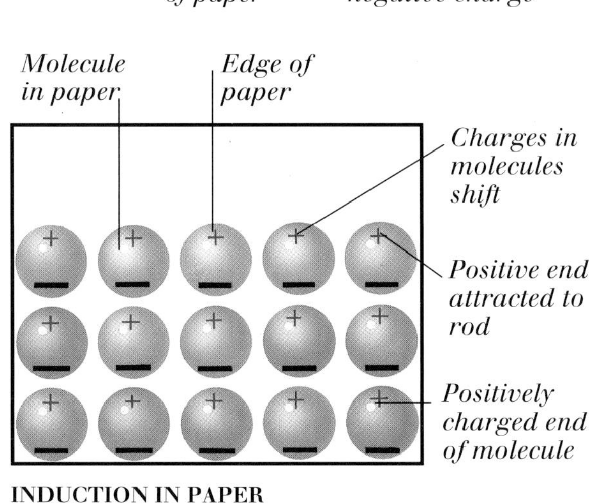

INDUCTION IN PAPER

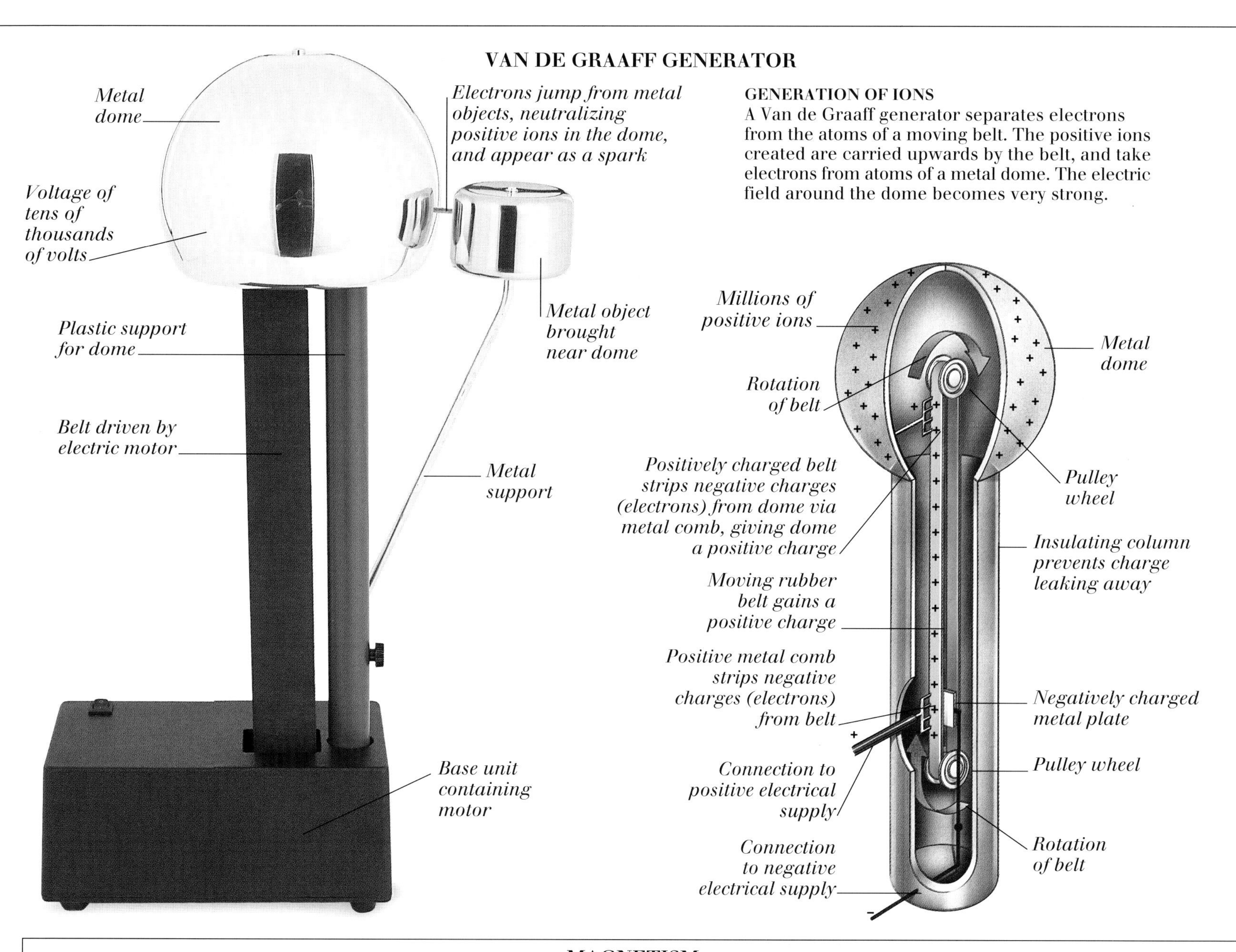

VAN DE GRAAFF GENERATOR
Metal dome
Voltage of tens of thousands of volts
Plastic support for dome
Belt driven by electric motor
Electrons jump from metal objects, neutralizing positive ions in the dome, and appear as a spark
Metal object brought near dome
Metal support
Base unit containing motor
GENERATION OF IONS
A Van de Graaff generator separates electrons from the atoms of a moving belt. The positive ions created are carried upwards by the belt, and take electrons from atoms of a metal dome. The electric field around the dome becomes very strong.
Millions of positive ions
Rotation of belt
Metal dome
Positively charged belt strips negative charges (electrons) from dome via metal comb, giving dome a positive charge
Pulley wheel
Insulating column prevents charge leaking away
Moving rubber belt gains a positive charge
Positive metal comb strips negative charges (electrons) from belt
Negatively charged metal plate
Pulley wheel
Connection to positive electrical supply
Connection to negative electrical supply
Rotation of belt

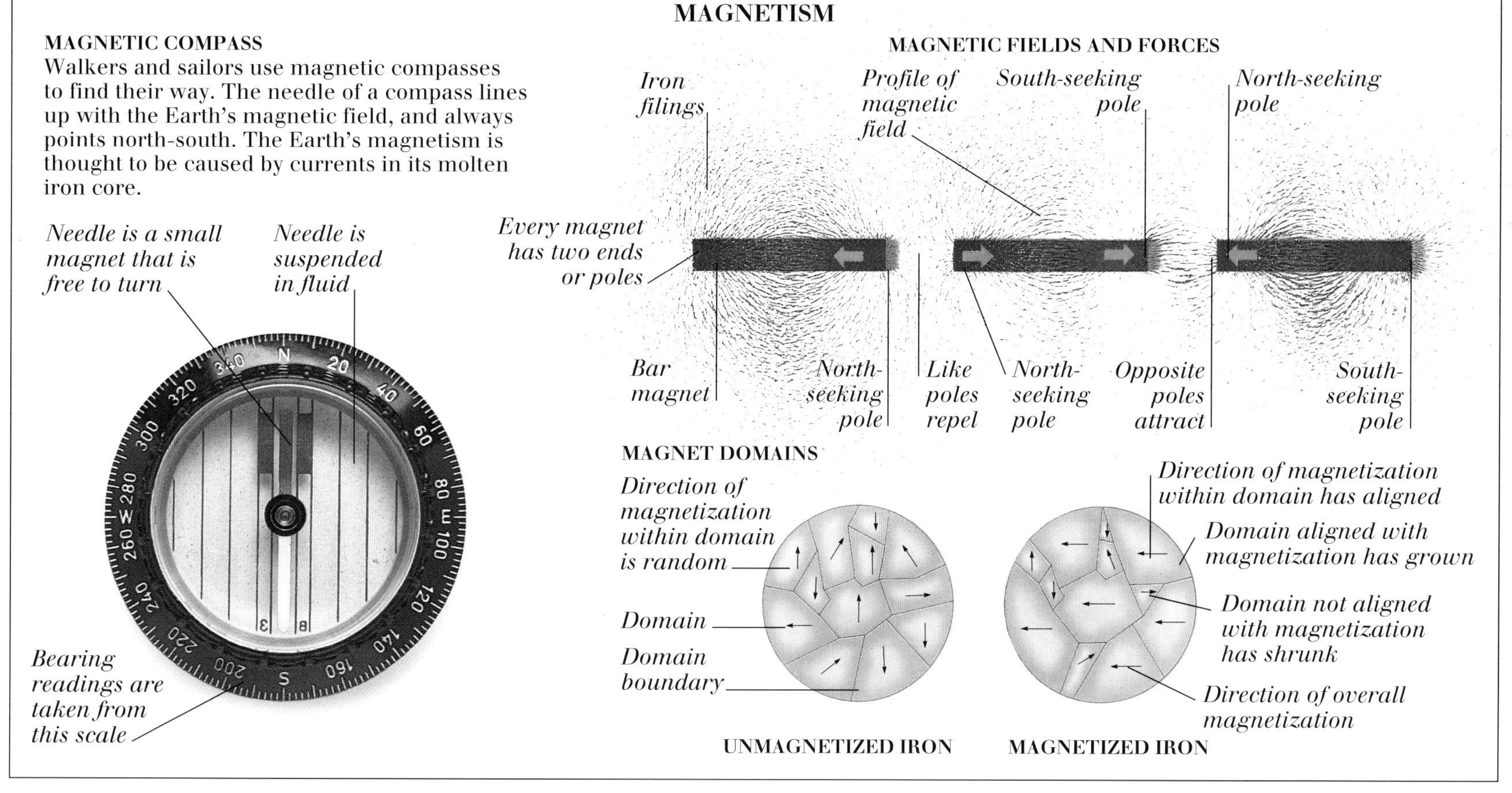

MAGNETISM
MAGNETIC COMPASS
Walkers and sailors use magnetic compasses to find their way. The needle of a compass lines up with the Earth's magnetic field, and always points north-south. The Earth's magnetism is thought to be caused by currents in its molten iron core.
Needle is a small magnet that is free to turn
Needle is suspended in fluid
Bearing readings are taken from this scale
MAGNETIC FIELDS AND FORCES
Iron filings
Profile of magnetic field
South-seeking pole
North-seeking pole
Every magnet has two ends or poles
Bar magnet
North-seeking pole
Like poles repel
North-seeking pole
Opposite poles attract
South-seeking pole
MAGNET DOMAINS
Direction of magnetization within domain is random
Domain
Domain boundary
Direction of magnetization within domain has aligned
Domain aligned with magnetization has grown
Domain not aligned with magnetization has shrunk
Direction of overall magnetization
UNMAGNETIZED IRON
MAGNETIZED IRON

Electric circuits

AN ELECTRIC CIRCUIT IS SIMPLY THE COURSE along which an **electric current** flows. **Electrons** carry negative charge and can be moved around a circuit by **electrostatic forces** (see pp. 30-31). A circuit usually consists of a **conductive** material, such as a metal, where the electrons are held very loosely to their atoms, thus making movement possible. The strength of the electrostatic force is the **voltage,** and is measured in volts (V). The resulting movement of electric charge is called an electric current, and is measured in amps (A). The higher the voltage, the greater the current will be. But the current also depends on the thickness, length, temperature, and nature of the material that conducts it. The **resistance** of a material is the extent to which it opposes the flow of electric current, and is measured in ohms (Ω). Good conductors have a low resistance, which means that a small voltage will produce a large current. In batteries, the dissolving of a metal **electrode** causes the freeing of electrons, resulting in their movement to another electrode and the formation of a current.

ELECTRIC CURRENT

Regions of positive or negative charge, such as those at the terminals of a battery, force electrons through a conductor. The electrons move from negative charge towards positive. Originally, current was thought to flow from positive to negative. This is so-called "conventional current".

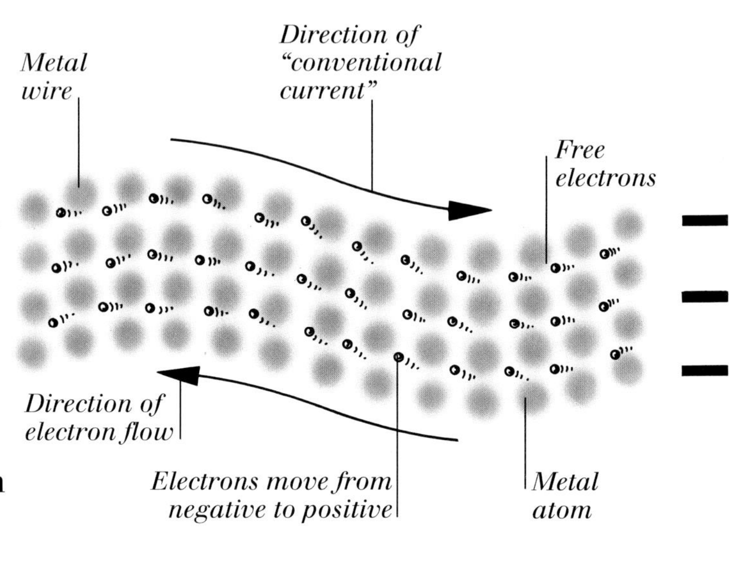

RESISTANCE

OHM'S LAW

A thin wire has a resistance to the flow of current. The longer and thinner the wire, the higher the resistance. An object's resistance can be worked out by dividing the voltage by the current (see p. 54).

22 Ω RESISTANCE

Electrical components called resistors allow current in circuits to be controlled. The current flowing around a circuit can be worked out using Ohm's Law.

47 Ω RESISTANCE

The larger the resistor, the smaller the current. The smaller the resistor, the larger the current.

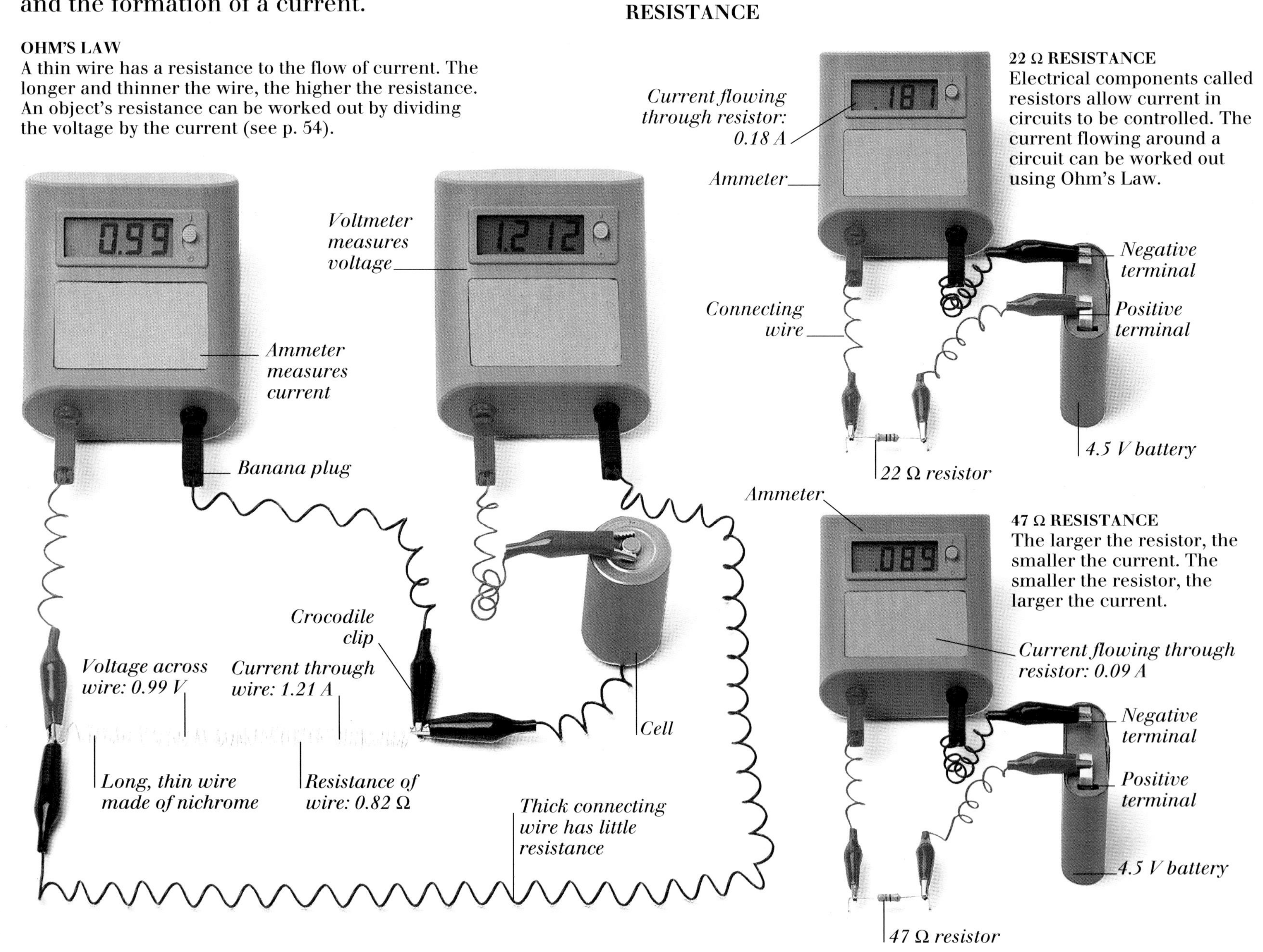

WORKING ELECTRIC CIRCUIT

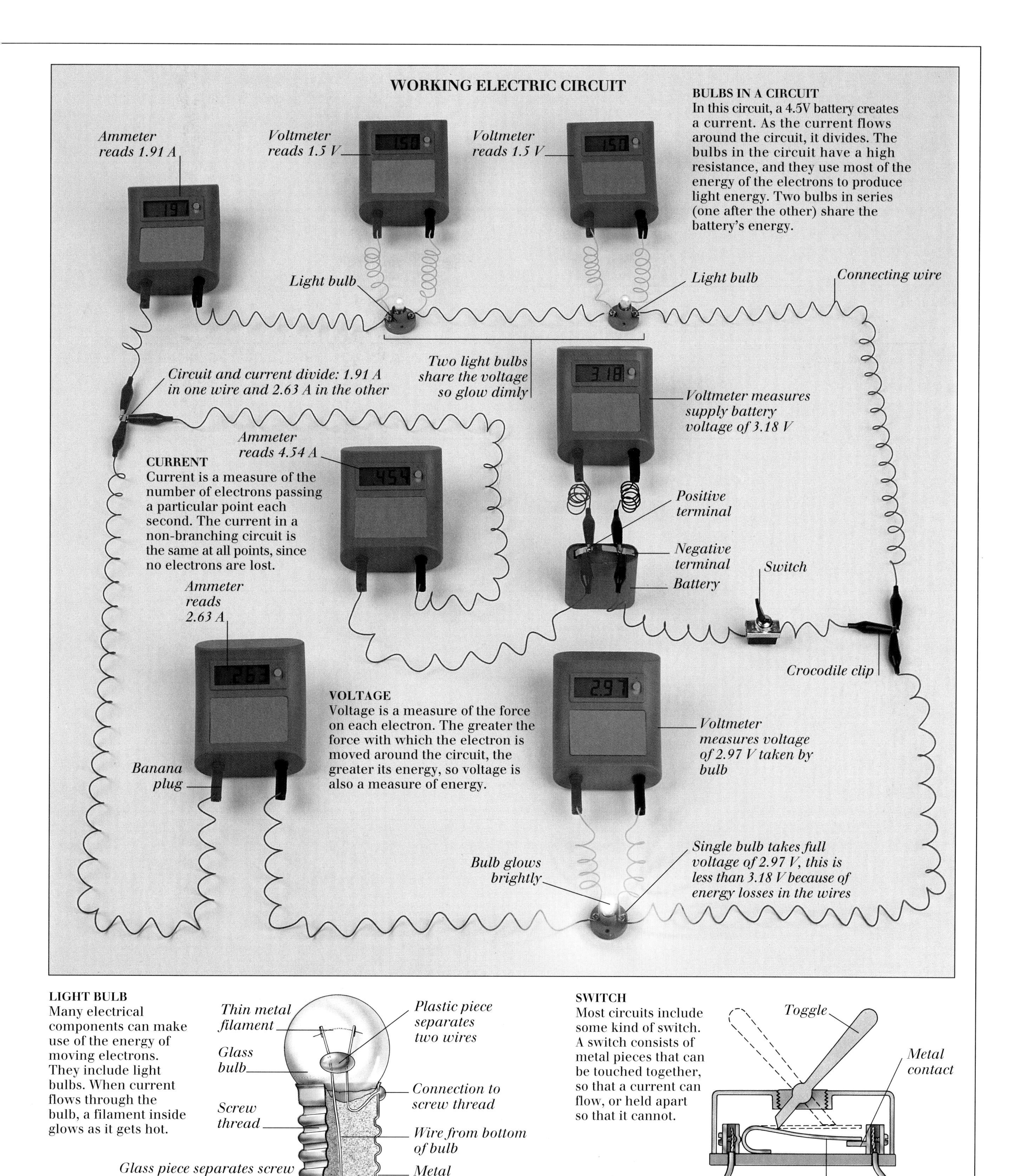

BULBS IN A CIRCUIT
In this circuit, a 4.5V battery creates a current. As the current flows around the circuit, it divides. The bulbs in the circuit have a high resistance, and they use most of the energy of the electrons to produce light energy. Two bulbs in series (one after the other) share the battery's energy.

CURRENT
Current is a measure of the number of electrons passing a particular point each second. The current in a non-branching circuit is the same at all points, since no electrons are lost.

VOLTAGE
Voltage is a measure of the force on each electron. The greater the force with which the electron is moved around the circuit, the greater its energy, so voltage is also a measure of energy.

LIGHT BULB
Many electrical components can make use of the energy of moving electrons. They include light bulbs. When current flows through the bulb, a filament inside glows as it gets hot.

SWITCH
Most circuits include some kind of switch. A switch consists of metal pieces that can be touched together, so that a current can flow, or held apart so that it cannot.

Electromagnetism

ANY ELECTRIC CURRENT WILL PRODUCE magnetism that affects iron filings and a compass needle in the same way as an ordinary, "permanent" magnet. The arrangement of "force lines" around a wire carrying an electric current – its **magnetic field** – is circular. The magnetic effect of electric current is increased by making the current-carrying wire into a coil. When a coil is wrapped around an iron bar, it is called an **electromagnet**. The magnetic field produced by the coil magnetizes the iron bar, strengthening the overall effect. A field like that of a bar magnet (see p. 31) is formed by the magnetic fields of the wires in the coil. The strength of the magnetism produced depends on the number of coils and the size of the current flowing in the wires. A huge number of machines and appliances exploit the connection between electricity and magnetism, including electric motors. Electromagnetic coils and permanent magnets are arranged inside an electric motor so that the forces of electromagnetism create rotation of a central spindle. This principle can be used on a large scale to generate immense forces.

MAGNETIC FIELD AROUND A CURRENT-CARRYING WIRE

The magnetic field produced by a current in a single wire is circular. Here, iron filings sprinkled around a current-carrying wire are made to line up by the magnetic field.

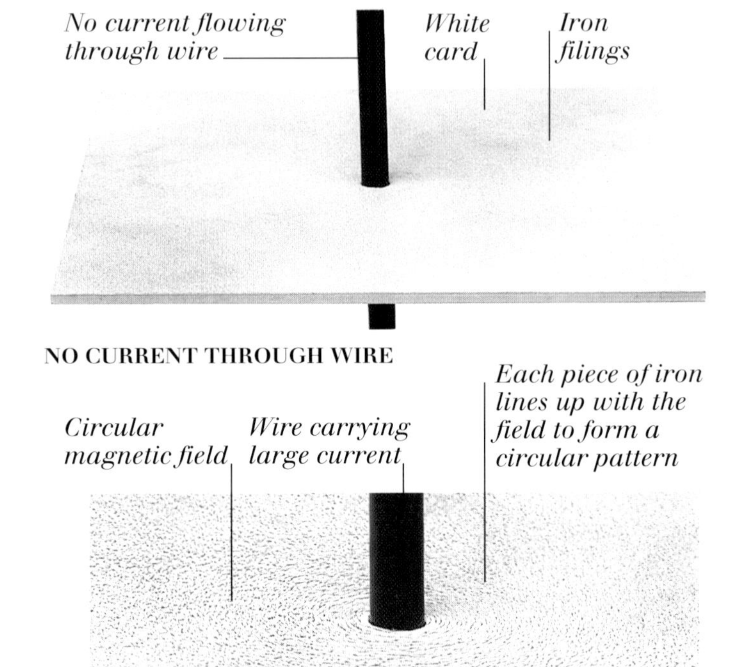

ELECTROMAGNETISM AFFECTING A COMPASS NEEDLE

A compass needle is a small magnet that is free to swivel around. It normally points north-south, in line with the Earth's magnetic field. But when a current flows in an adjacent wire, the needle swings around to line up with the field created by the current.

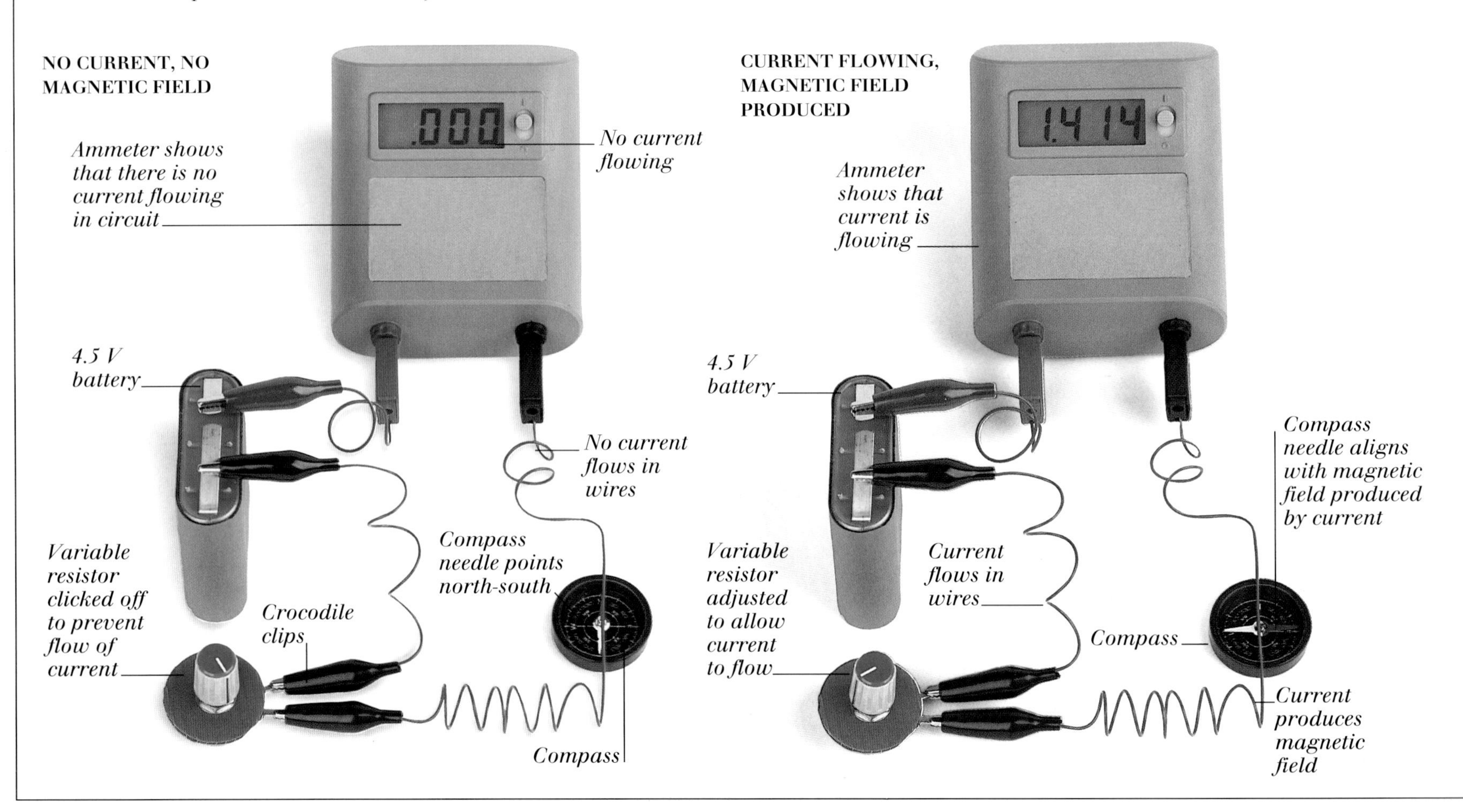

THE STRENGTH OF AN ELECTROMAGNET

An electromagnet is a coil of wire wrapped around an iron bar. It behaves like a permanent magnet, except that it can be turned off. Here, the size of the magnetic force produced by an electromagnet is measured by the number of paper clips it can lift. The strength of an electromagnet depends on the number of turns in the coil and the current flowing through the wire.

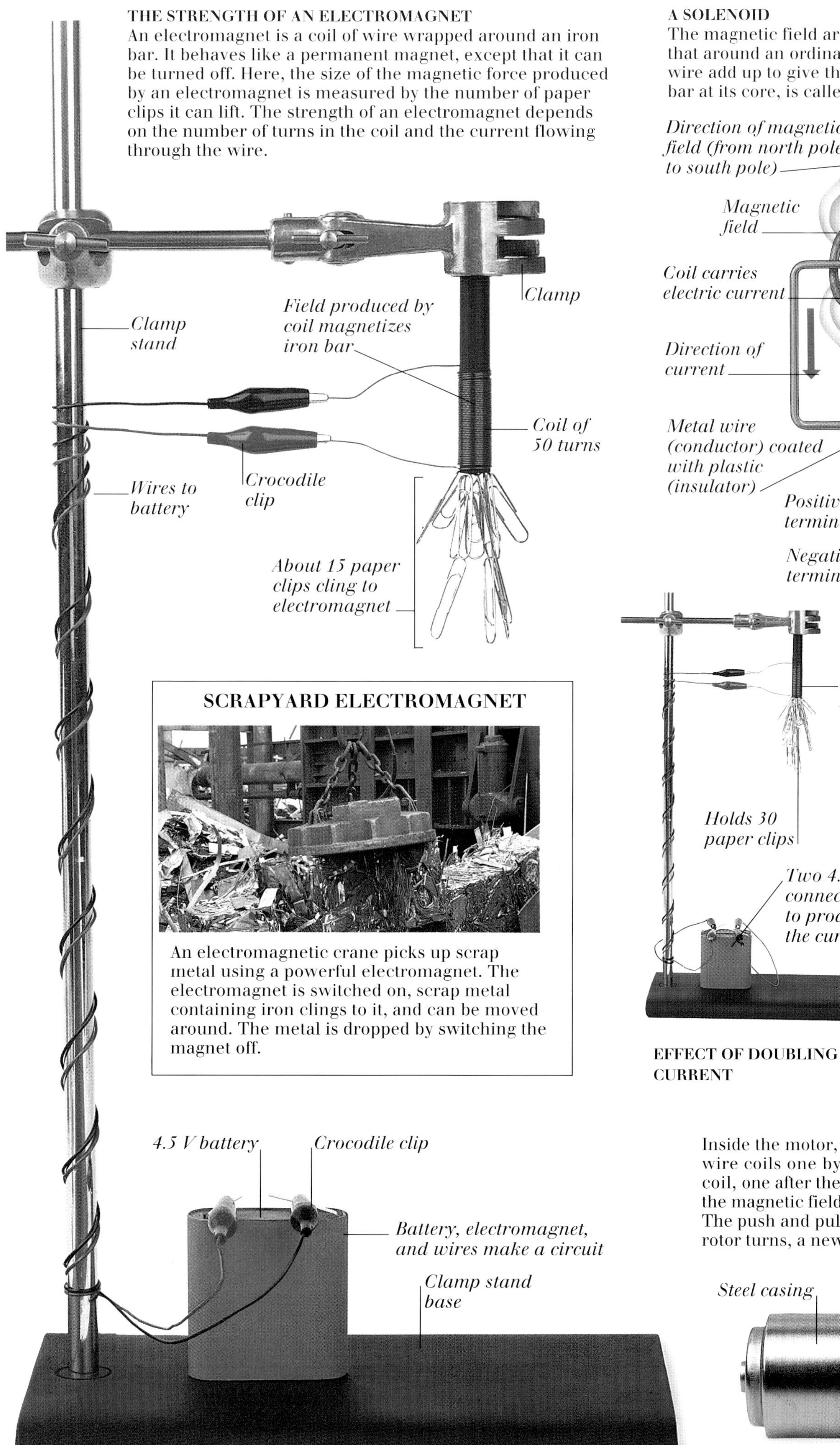

SCRAPYARD ELECTROMAGNET

An electromagnetic crane picks up scrap metal using a powerful electromagnet. The electromagnet is switched on, scrap metal containing iron clings to it, and can be moved around. The metal is dropped by switching the magnet off.

A SOLENOID

The magnetic field around a coil of current-carrying wire resembles that around an ordinary bar magnet. The fields of each individual wire add up to give the overall pattern. A coil like this, with no iron bar at its core, is called a **solenoid**.

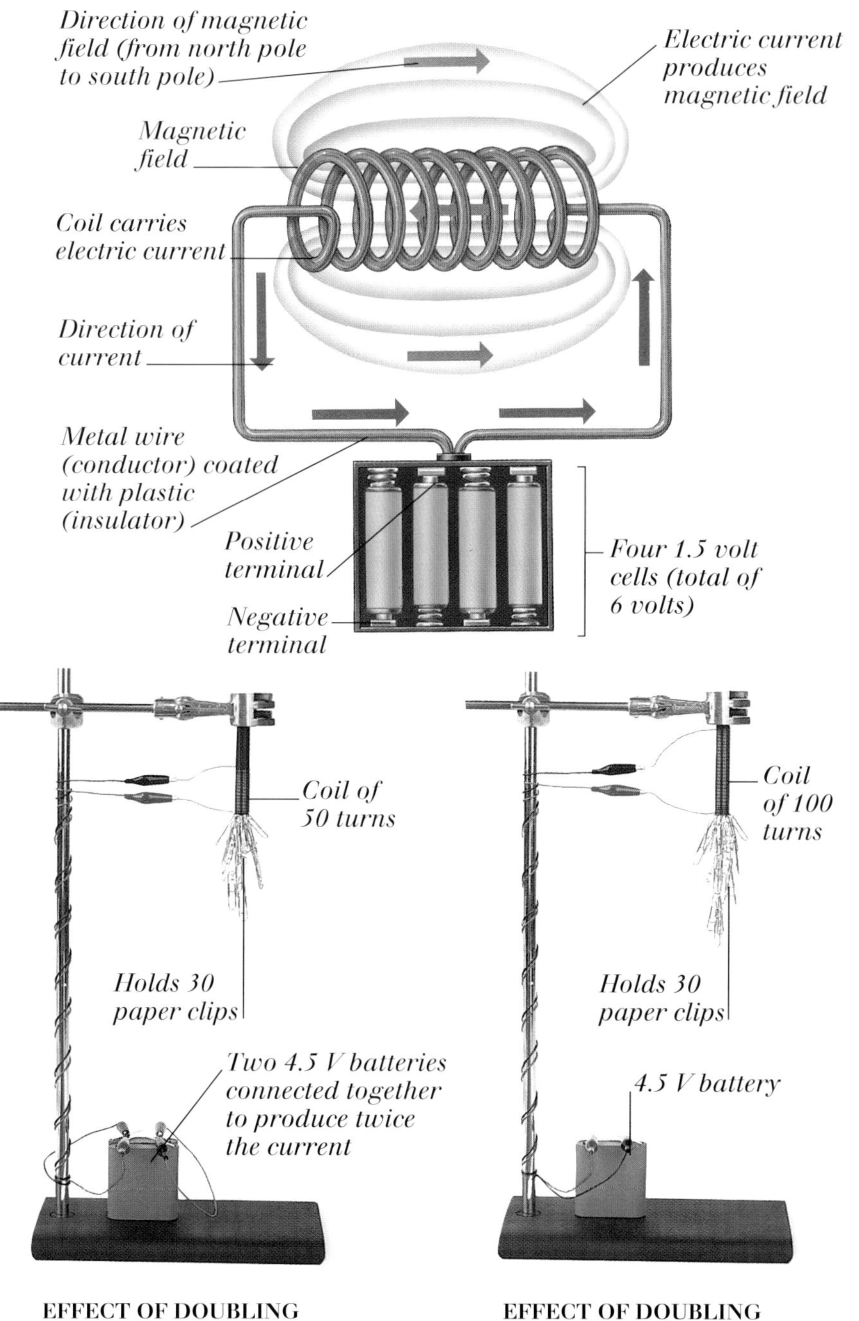

EFFECT OF DOUBLING CURRENT

EFFECT OF DOUBLING NUMBER OF TURNS ON COIL

ELECTRIC MOTORS

Inside the motor, an electric current is sent through a series of wire coils one by one, providing a magnetic field around each coil, one after the other. The magnetism of the coils interacts with the magnetic fields of permanent magnets placed around them. The push and pull of this interaction turns the motor. As the rotor turns, a new coil is activated and the motion continues.

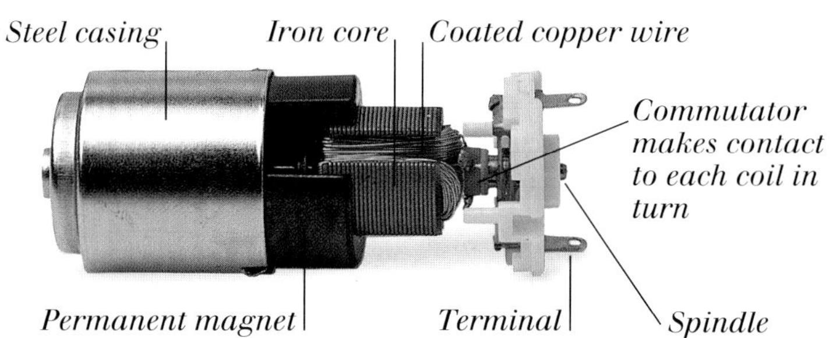

Generating electricity

THERE ARE MANY WAYS TO GENERATE electricity. The most common is to use coils of wire and magnets in a **generator**. Whenever a wire and magnet are moved relative to each other, a **voltage** is produced. In a generator, the wire is wound into a coil. The more turns in the coil and the faster the coil moves, the greater the voltage. The coils or magnets spin around at high speed, turned by water pressure, the wind, or, most commonly, by steam pressure. The steam is usually generated by burning coal or oil, a process that creates pollution. Renewable sources of electricity – such as hydroelectric power, wind power, solar energy, and geothermal power – produce only heat as pollution. In a generator, the **kinetic energy** of a spinning object is converted into electrical energy. A solar cell converts the energy of sunlight directly into electrical energy, using layers of **semiconductors.**

GENERATOR

Inside a generator, you will find coils of wire and magnets (or electromagnets). In the generator shown below, electromagnets spin rapidly inside stationary coils of wire. A voltage is then produced in the coils.

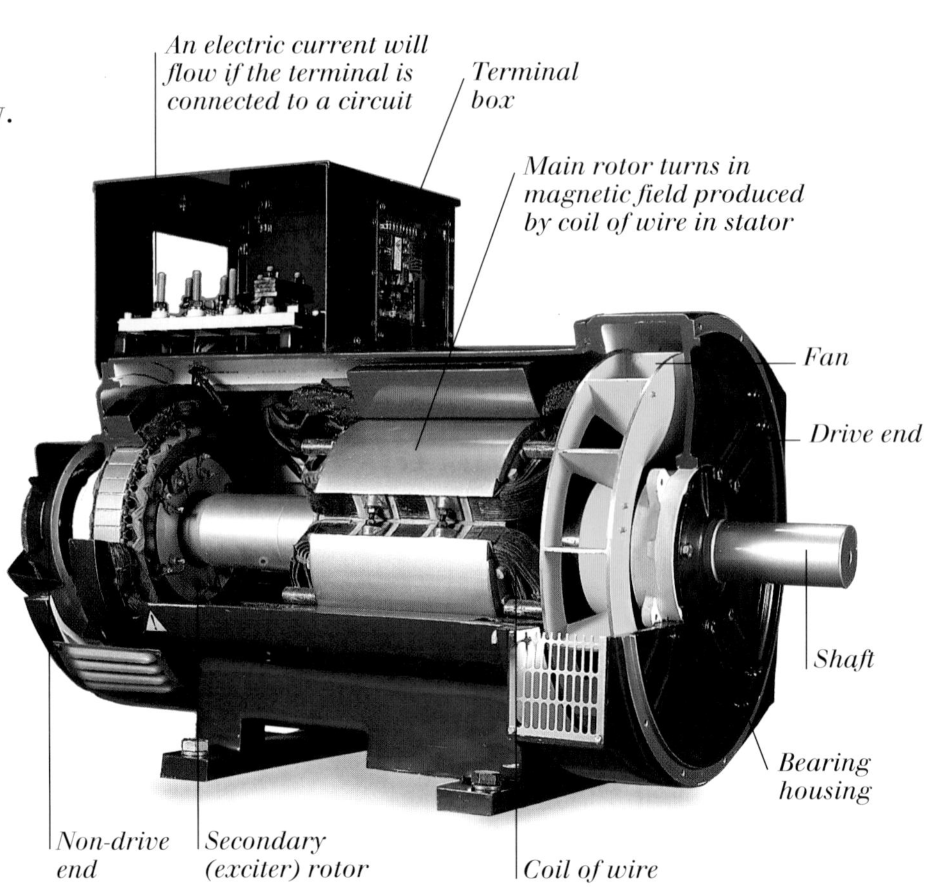

WATER POWER

HYDROELECTRIC POWER STATION
Water flows into a hydroelectric power station from a reservoir above. The water exerts pressure on **turbines** within the power station. The pressure pushes the water through the turbines, turning them at great speed. The turbine runs a generator, which produces the electricity.

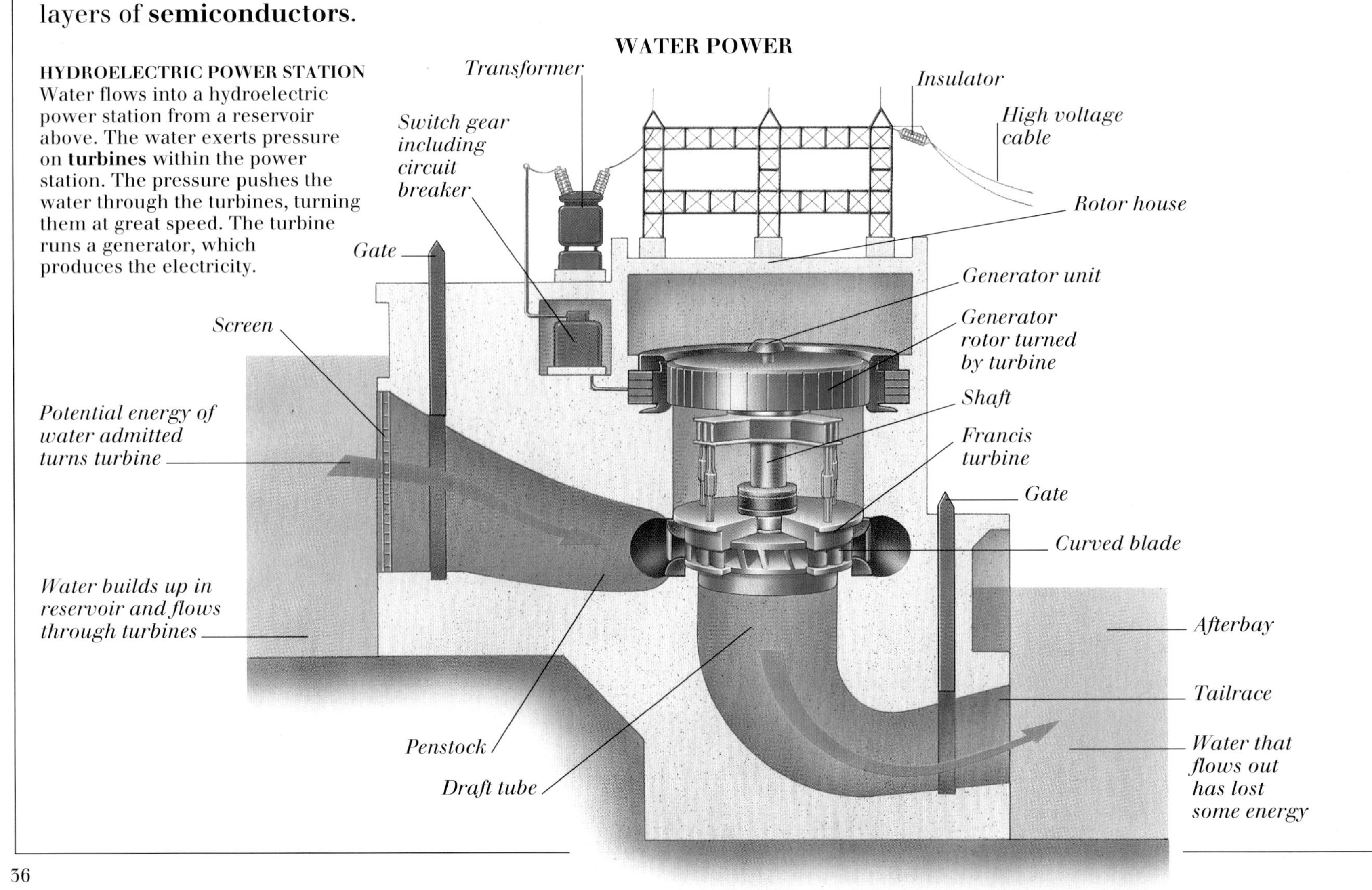

WIND POWER

WIND TURBINE

Energy from the wind is converted to electricity by wind turbines. The rotating turbine blades are connected to a generator, which produces a voltage. The faster the wind blows and the larger the blades, the greater the energy available.

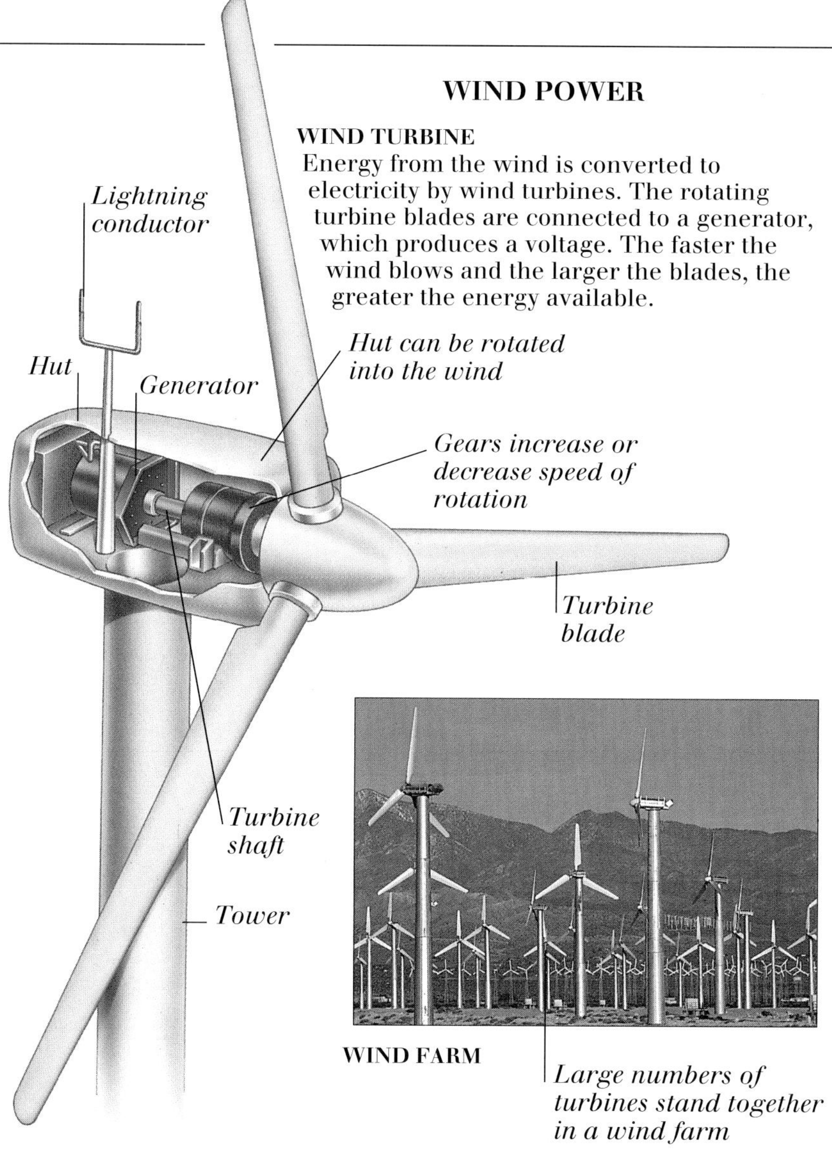

WIND FARM

OTHER SOURCES

Two further examples of renewable sources are tidal power and geothermal power. The tides are a result of the gravitational pull of the Moon. Geothermal heat is produced by the disintegration of radioactive atoms in the Earth's core.

GEOTHERMAL POWER

Water pumped underground is turned into high-pressure steam by geothermal heat. The steam returns to the surface under pressure and turns turbines.

TIDAL POWER STATION

Sea water is held back by a barrage as it rises and falls. When there is a difference in height between the water on either side of the barrage, the water escapes through tunnels, turning turbines.

SOLAR ENERGY

The energy of sunlight produces electricity in solar cells by causing **electrons** to leave the **atoms** in a semiconductor. Each electron leaves behind a gap, or **hole**. Other electrons move into the hole, leaving holes in their atoms. This process continues all the way around a circuit. The moving chain of electrons is an electric current.

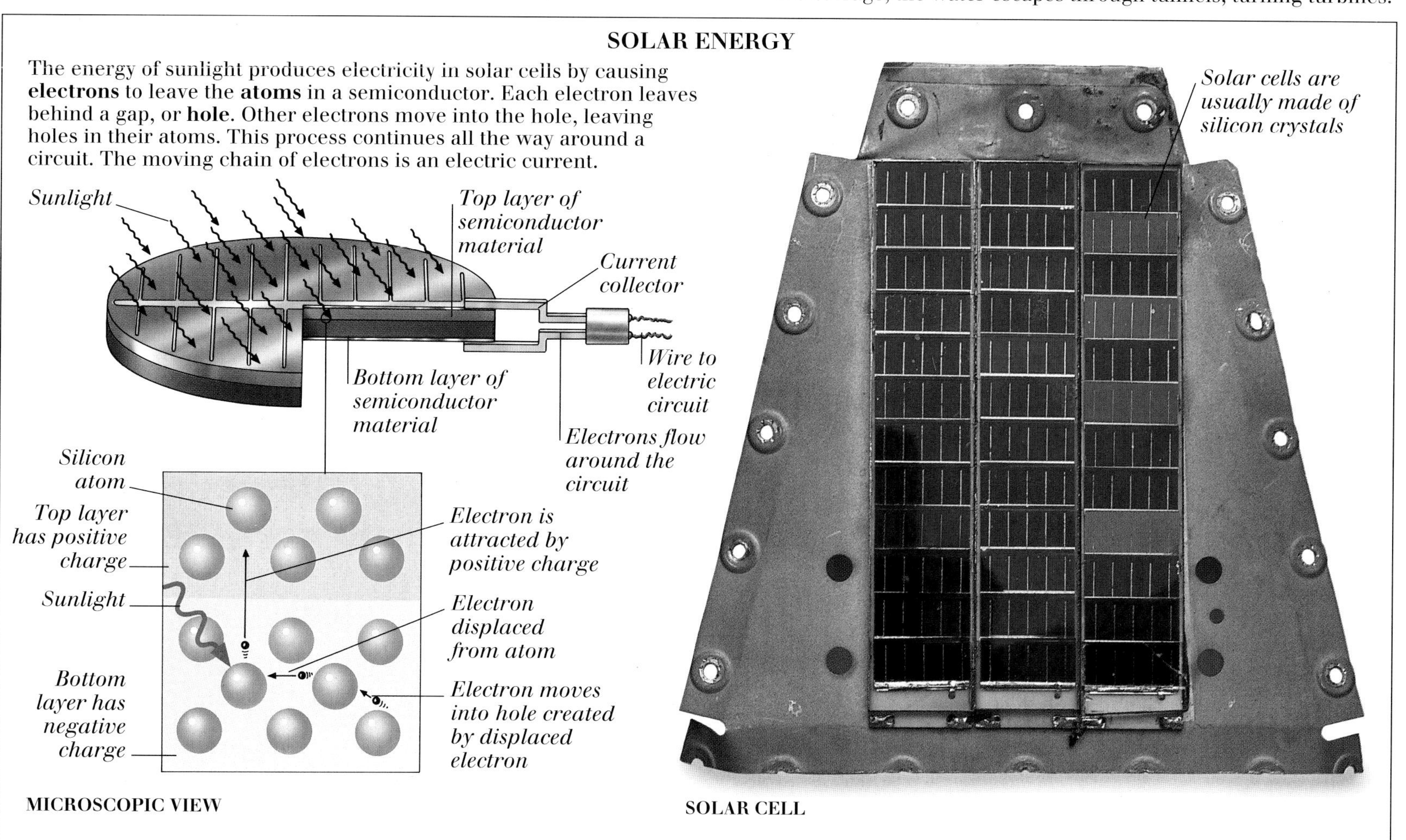

MICROSCOPIC VIEW

SOLAR CELL

Electromagnetic radiation

ELECTRICITY AND MAGNETISM ARE DIRECTLY related (see pp. 34-37): a changing **electric field** will produce a changing **magnetic field**, and vice versa. Whenever an electric charge, such as that carried by an **electron**, accelerates, it gives out energy in the form of electromagnetic radiation. For example, electrons moving up and down a radio antenna produce a type of radiation known as radio waves. Electromagnetic radiation consists of oscillating electric and magnetic fields. There is a wide range of different types of electromagnetic radiation, called the electromagnetic spectrum, extending from low-energy radio waves to high-energy, short-**wavelength** gamma rays. This includes visible light and X-rays. Electromagnetic radiation can be seen as both a wave motion (see pp. 20-21) or as a stream of particles called **photons** (see pp. 48-49). Both interpretations are useful, as they each provide a means for predicting the behaviour of electromagnetic radiation.

RADIATION AS PARTICLES AND WAVES

OSCILLATING FIELDS

All electromagnetic radiation has behaviour typical of waves, such as **diffraction** and **interference**. It can be thought of as a combination of changing electric and magnetic fields.

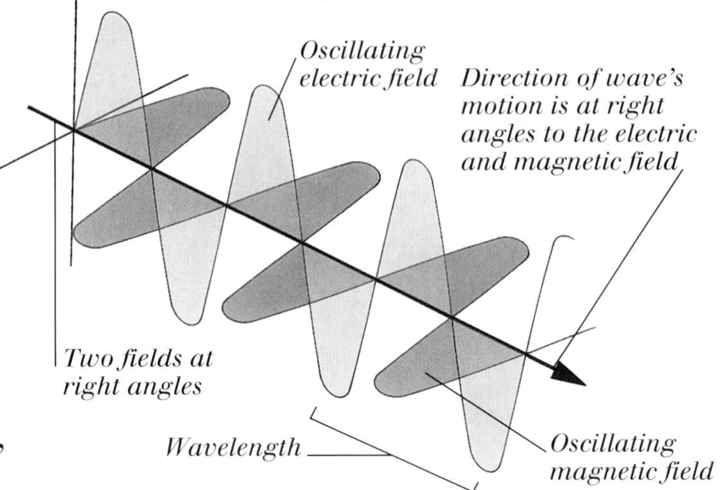

PHOTONS

All electromagnetic radiation also has behaviour typical of particles. For example, its energy comes in individual bundles called photons.

RADIO WAVES

PRODUCTION OF RADIO WAVES

The **electric current** in a radio antenna changes direction rapidly, and produces a changing magnetic field around the antenna. This magnetic field produces an electric field, which in turn produces a magnetic field, and so on.

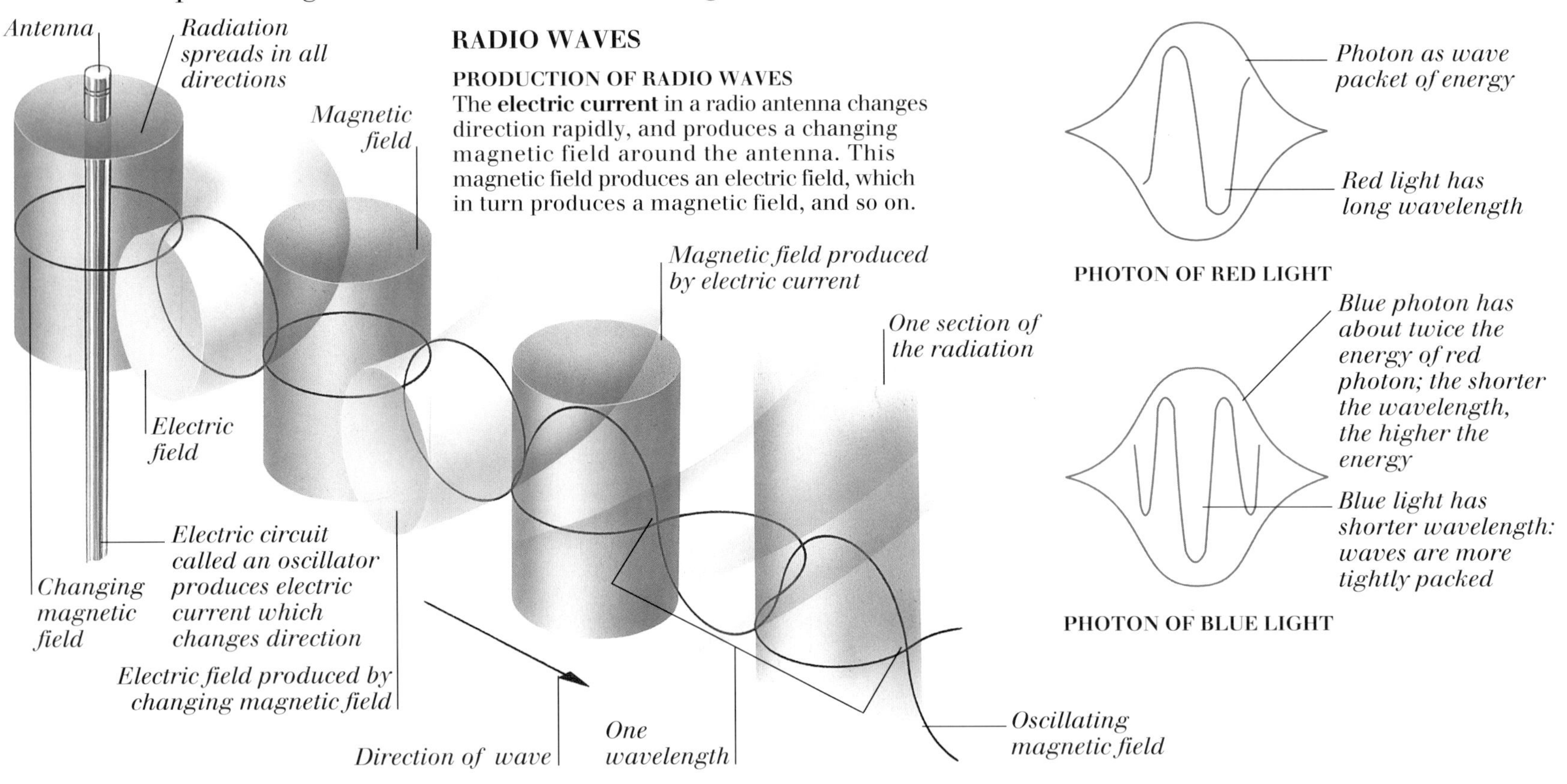

THE ELECTROMAGNETIC SPECTRUM

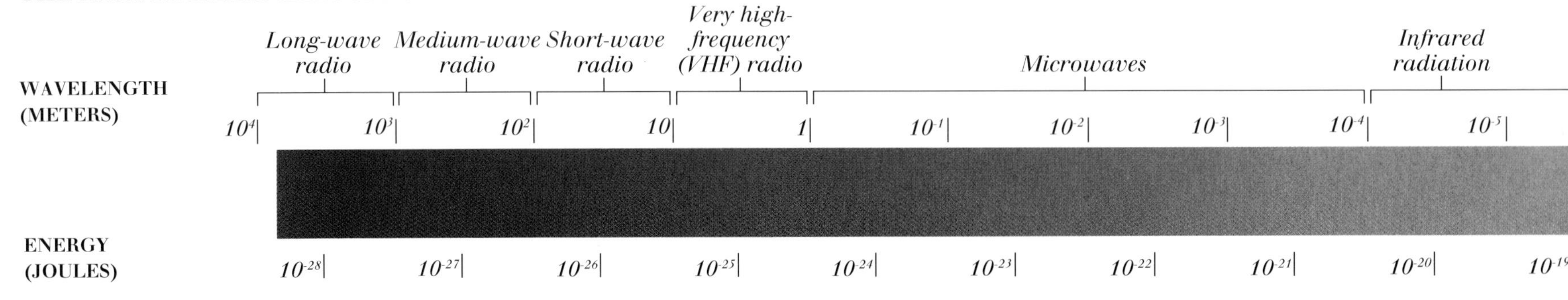

THE WHITE LIGHT SPECTRUM

Human eyes can detect a range of wavelengths of electromagnetic radiation, from "red light" to "blue light". When all of the wavelengths within that range are perceived together, they produce the sensation of white light.

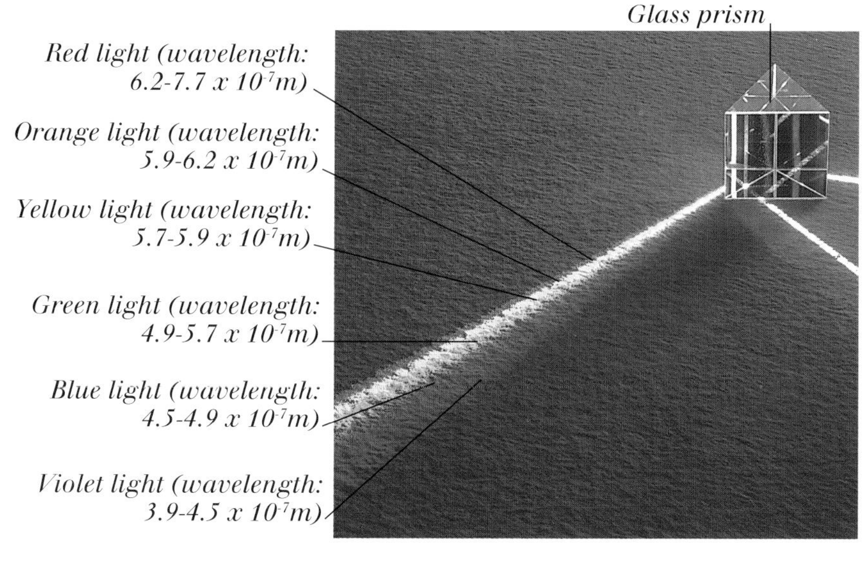

X-RAYS

PRODUCTION OF X-RAYS

Near the high-energy end of the electromagnetic spectrum come X-rays. In an X-ray tube, electrons are accelerated by a strong electric field. They then hit a metal target, and their kinetic energy is turned into electromagnetic radiation.

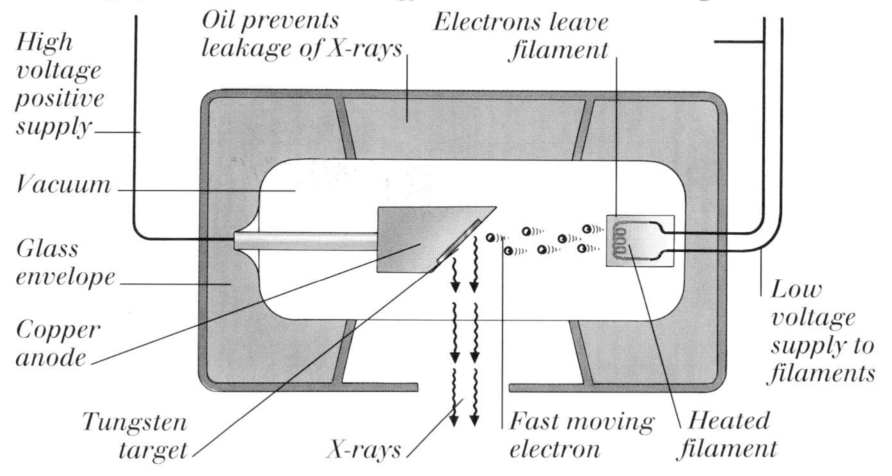

X-RAY PHOTOGRAPH

The main use for X-rays is in medical photography. Radiation from an X-ray tube does not pass through bone, so when an image is recorded on paper sensitive to X-rays, an image of the bone remains. Thus fractures can be investigated without the need for surgery.

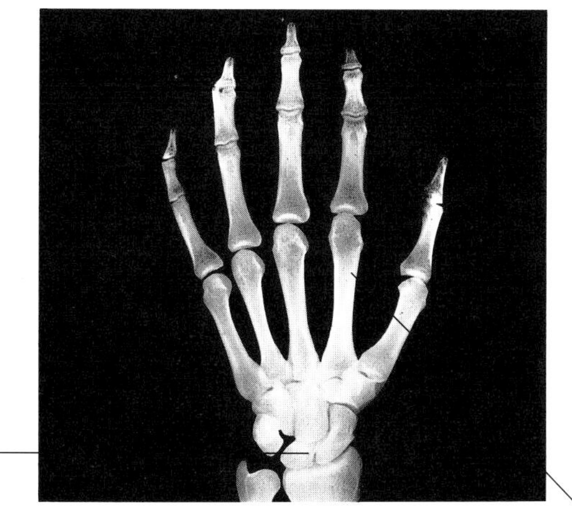

RADIATION FROM HOT OBJECTS

The **atoms** of a solid vibrate (see pp. 22-23). Atoms contain electric charges in the form of **protons** and electrons. Because they vibrate, these charges produce a range of electromagnetic radiation. The rate of vibration, and therefore the wavelengths of radiation produced, depends on **temperature,** as this steel bar shows.

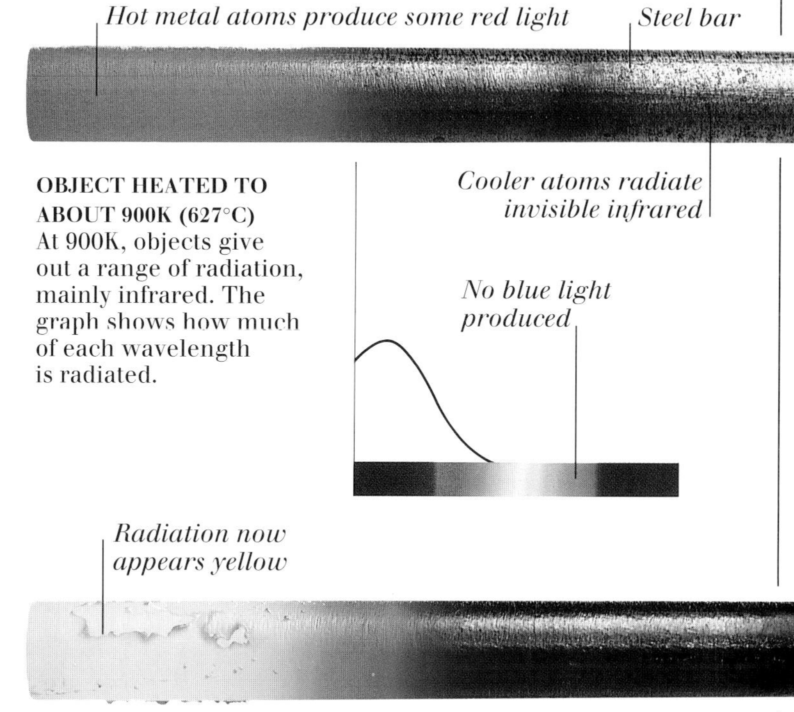

OBJECT HEATED TO ABOUT 900K (627°C)
At 900K, objects give out a range of radiation, mainly infrared. The graph shows how much of each wavelength is radiated.

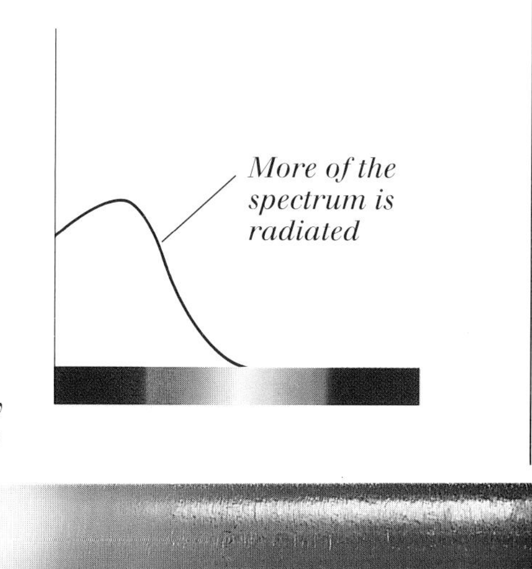

OBJECT HEATED TO ABOUT 1,500K (1,227°C)
As the metal atoms vibrate more vigorously, the radiation has more energy. It therefore includes more of the visible spectrum.

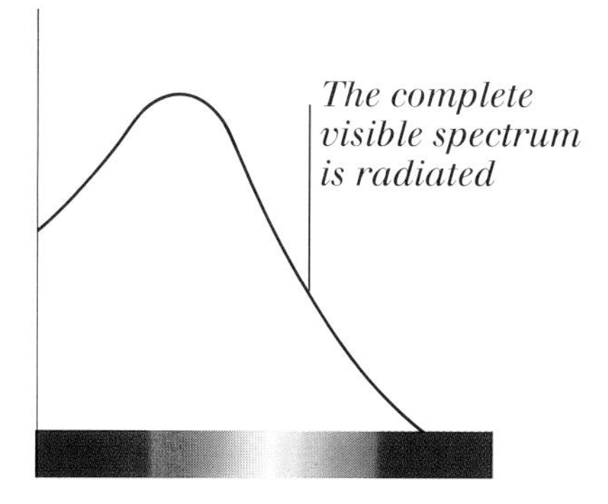

OBJECT HEATED TO ABOUT 1,800K (1,527°C)
Near its melting point, the bar produces even more light. The range of light now includes the entire visible spectrum. This is why it looks bright white.

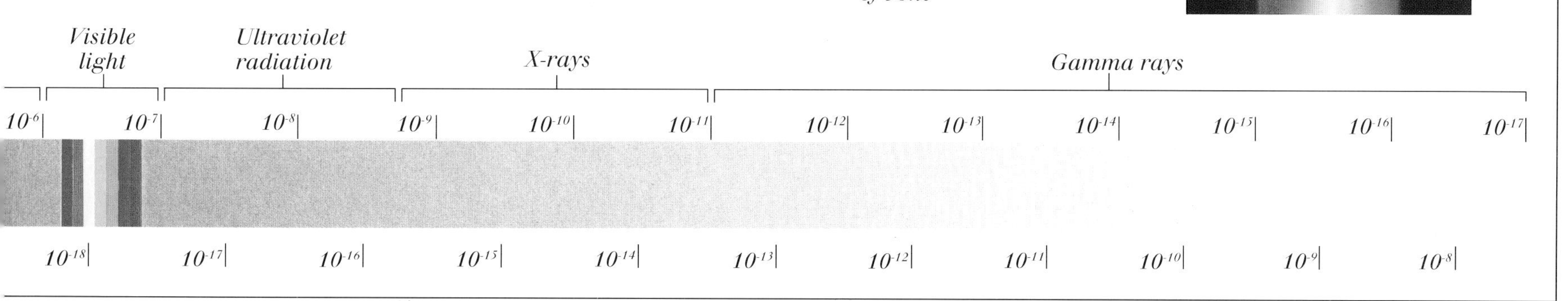

Colour

THE HUMAN EYE CAN ONLY PERCEIVE a small section of the **electromagnetic spectrum** (see pp. 38-39). We call this section “visible light”. Different colours across the spectrum of visible light correspond to different **wavelengths** of light. Our eyes contain cells called **cones**, which are sensitive to these different wavelengths and allow us to see in colour. Three different types of cone are affected by light in the red, green, and blue parts of the spectrum. These correspond to the **primary colours**. Different light sources give out different parts of the spectrum, which appear as different colours. When combined, coloured lights appear as different colours. This is called the **additive process**. Adding primary light sources in the correct proportions can produce the sensation of other colours in our eyes. When light hits a **pigment** in an object, only some colours are reflected. Which colours are **reflected** and which **absorbed** depends on the pigment. This is the **subtractive process**. Looking at a coloured object in coloured light may make it appear different. This is because pigments can only reflect colours that are present in the incoming light.

CONE SENSITIVITY

Sensitivity of green cone peaks in the green part of the spectrum

Sensitivity of blue cone peaks in the blue part of the spectrum

Sensitivity of red cone peaks in the red part of the spectrum

Red and blue sensitivity does not overlap

White light (visible) spectrum

COLOUR VISION
There are three different types of cone in the normal human eye, each sensitive to a different part of the spectrum. White light stimulates all three types of cone cells.

SOURCES OF LIGHT

BRIGHT FILAMENT LAMP

This spectrum shows which colours are produced

All colours of light together combine to produce white

BRIGHT FILAMENT LAMP
With a high **electric current**, the whole spectrum of visible light is produced (see p. 39).

LED produces colours in the green part of the spectrum

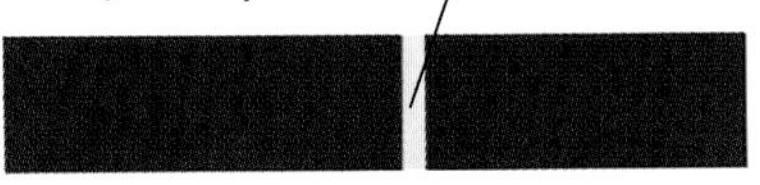

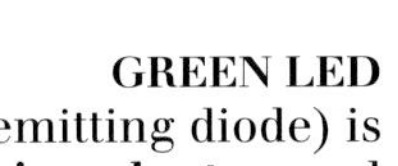

LED appears green

GREEN LED
An LED (light-emitting diode) is made of a **semiconductor**, and produces certain colours of light.

GREEN LED

DIM FILAMENT LAMP

Red, yellow, and green light combine to produce orange

Lamp appears orange

No blue light produced

DIM FILAMENT LAMP
With a smaller current, the **temperature** of the **filament** (see pp. 32-33) is low.

Two colours of light very close together in the orange part of the spectrum are produced

Lamp appears orange

SODIUM LAMP
In a sodium lamp, an electric current excites electrons in sodium vapour, giving them extra energy. The electrons give the energy out as light.

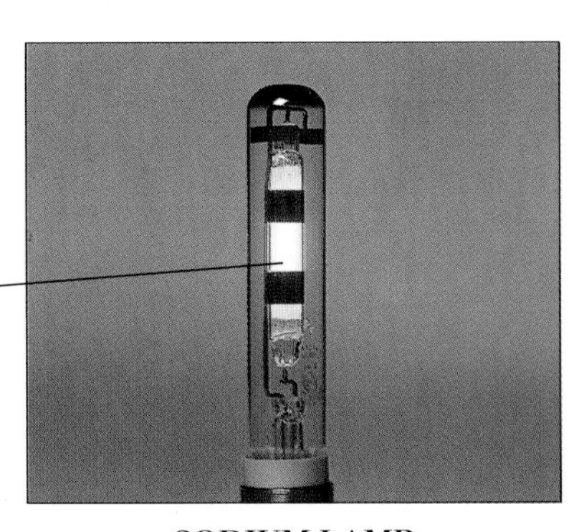

SODIUM LAMP

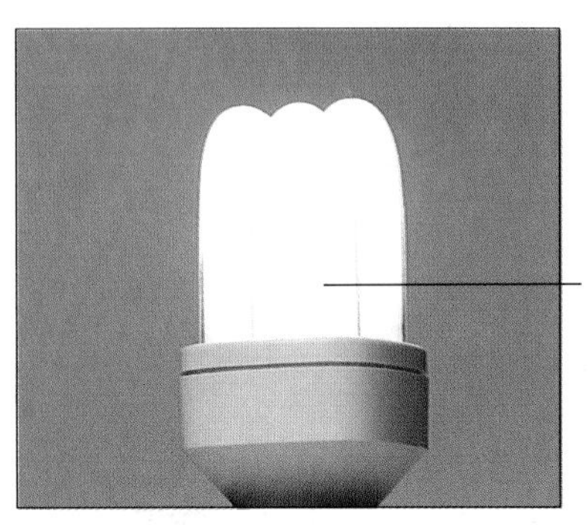

FLUORESCENT LAMP

Lamp produces certain colours in each part of the spectrum

All three types of cone are stimulated and lamp appears white

FLUORESCENT LAMP
In a **fluorescent** lamp, chemicals called **phosphors** produce colours in many parts of the spectrum.

Only certain colours characteristic of neon are produced

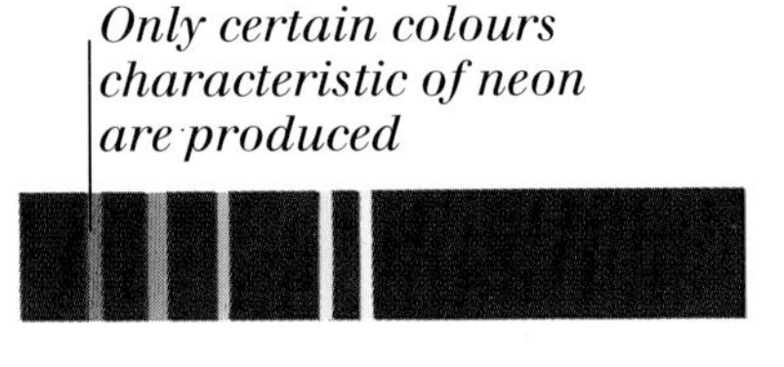

Lamp appears orange

NEON TUBE
In a similar way to a sodium lamp, a neon discharge lamp produces a characteristic orange glow.

NEON TUBE

ADDITIVE PROCESS

Adding red, green, and blue light in the correct proportions can create the illusion of any other colour. These three colours are called primary colours. A colour made from adding any two primary colours alone is called a secondary colour.

MAGENTA (SECONDARY)

Primary red and primary blue combine to appear as magenta

WHITE LIGHT

All the primary colours together stimulate all types of cone and appear white

RED LIGHT (PRIMARY)

Primary red light stimulates the red cone

PRIMARY COLOURS FOR THE ADDITIVE PROCESS

BLUE LIGHT (PRIMARY)

Primary blue light stimulates the blue cone

CYAN (SECONDARY)

Primary green and primary blue combine to appear as cyan

GREEN LIGHT (PRIMARY)

Primary green light stimulates the green cone

YELLOW (SECONDARY)

Primary red and primary green combine to appear as yellow

SUBTRACTIVE PROCESS

These three filters contain pigments which absorb some of the colours in the white light passing through them from a light beneath. By mixing primary pigments together, all colours except true white can be produced.

The primary pigment colours are different to the primary light colours

CYAN FILTER (PRIMARY)

A primary cyan filter will absorb all light except blue and green

BLUE (SECONDARY)

Magenta and cyan filters together only allow blue light through

GREEN (SECONDARY)

Cyan and yellow filters together only allow green light through

BLACK (NO COLOUR)

Where all three filters overlap, they absorb all colours, and appear black

YELLOW FILTER (PRIMARY)

A primary yellow filter will absorb all light except red and green

MAGENTA FILTER (PRIMARY)

A primary magenta filter will absorb all light except red and blue

RED (SECONDARY)

Magenta and yellow filters together only allow red light through

COMBINING PRIMARY COLOURED FILTERS FOR THE SUBTRACTIVE PROCESS

COLOURED OBJECTS IN COLOURED LIGHT

Green pot appears green

White pot reflects all colours

IN WHITE LIGHT

The green pot only reflects the green part of the spectrum, absorbing the other colours.

Green pot appears black

White pot reflects the blue light and appears blue

IN BLUE LIGHT

When only blue light is available, the green pigment can reflect no green light, and appears black.

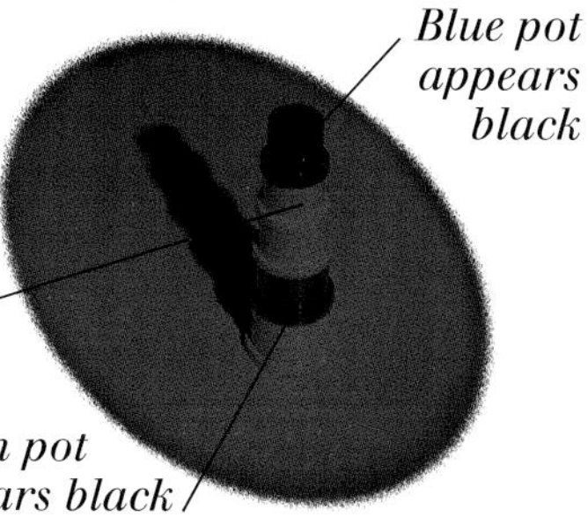

Blue pot appears black

Red pot appears red

Green pot appears black

IN RED LIGHT

When only red light is available, the green pigment can reflect no green light, and appears black.

Red pot appears black

Blue pot appears black

White pot appears green

IN GREEN LIGHT

When only green light is available, the green pigment reflects green light and appears green.

Reflection and refraction

LIGHT IS A FORM OF **electromagnetic radiation** (see pp. 48-49). In free space, it travels in a straight line at 300 million metres per second. When a beam of light meets an object, a proportion of the rays may be reflected. Some light may also be absorbed, and some transmitted. Without reflection, we would only be able to see objects that give out their own light. Light always reflects from a surface at the same angle at which it strikes it. Thus parallel rays of light meeting a very flat surface will remain parallel when reflected. A beam of light reflecting from an irregular surface will scatter in all directions. Light that passes through an object will be **refracted**, or bent. The angle of refraction depends on the angle at which the light meets the object, and the material from which the object is made. **Lenses** and mirrors can cause light rays to **diverge** or **converge**. When light rays converge, they can reach a point of focus. For this reason, lenses and mirrors can form images. This is useful in binoculars and other optical instruments (see pp. 44-45).

SEEING BY REFLECTED LIGHT

Light travels out from a source and hits objects such as this plant. The plant reflects some of this light, a proportion of which will enter our eyes.

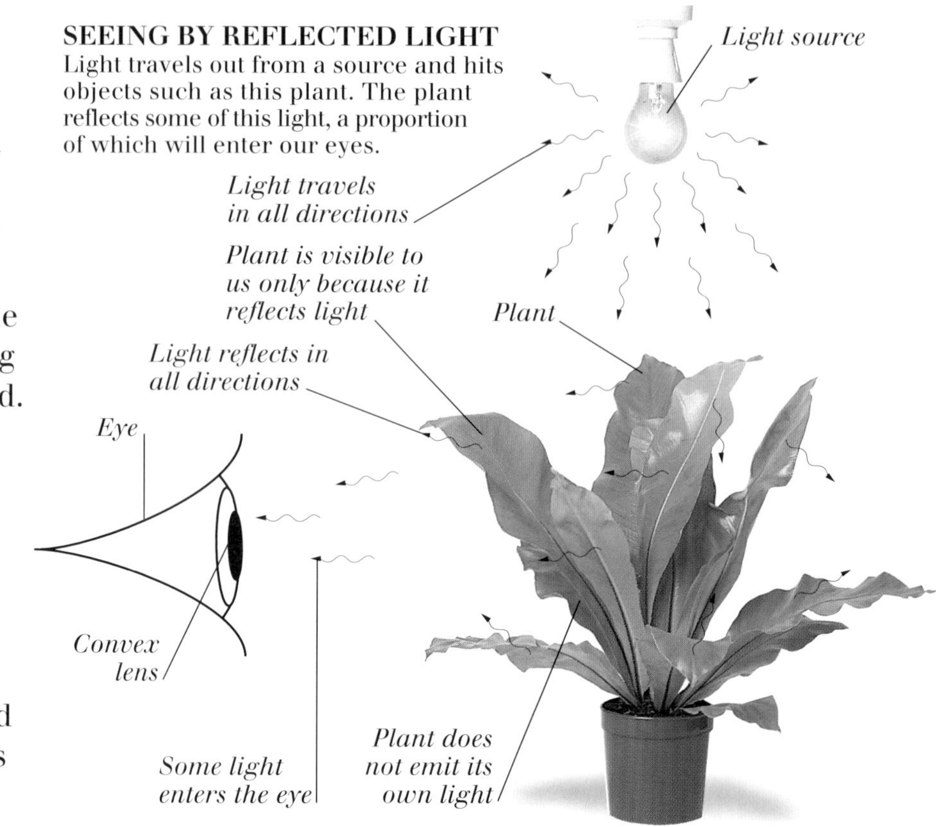

REFLECTING AND REFRACTING

The illustrations below show what happens when parallel beams of light reflect regularly and irregularly and when they refract.

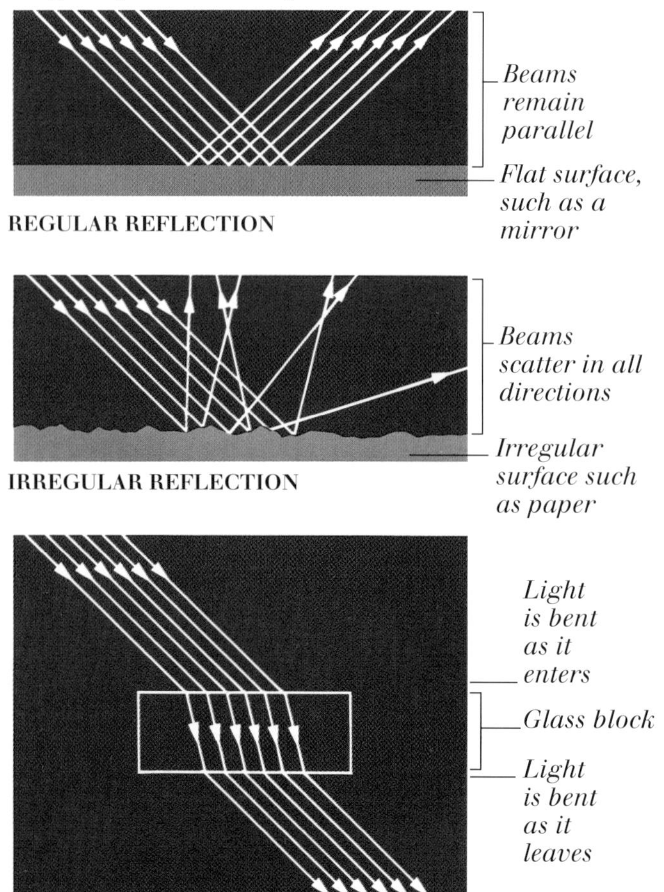

REGULAR REFLECTION

IRREGULAR REFLECTION

REFRACTION IN A GLASS BLOCK

TOTAL INTERNAL REFLECTION

When light moves from one medium to another, for example from glass to air, some of the light will normally be reflected. When the light striking the boundary reaches a certain angle – the **critical angle** – all of the light reflects back. This is called **total internal reflection**. It is put to use in binoculars, where the light path is folded by prisms so that it can be contained within a compact case.

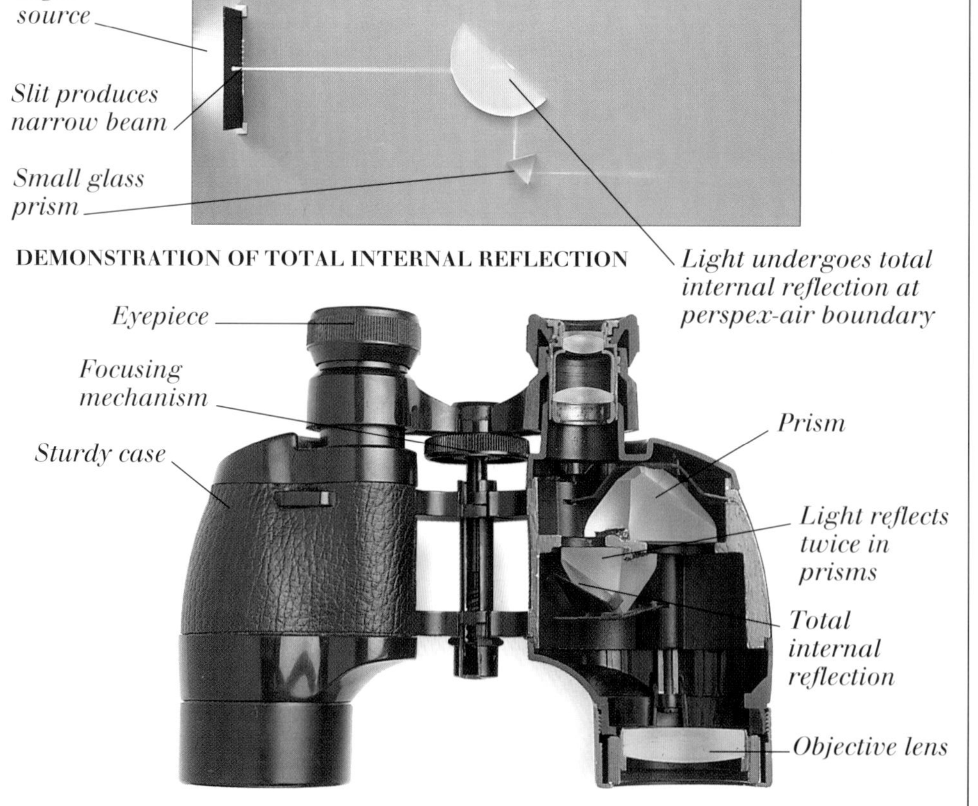

DEMONSTRATION OF TOTAL INTERNAL REFLECTION

BINOCULARS

LENSES AND MIRRORS

The images below show how beams of light from a bulb are affected by **concave** and **convex** mirrors and lenses. Convex lenses and mirrors have surfaces that curve outwards at the centre, while concave lenses curve inwards and are thicker at the edges.

CONCAVE LENS (BENDS LIGHT OUTWARDS)

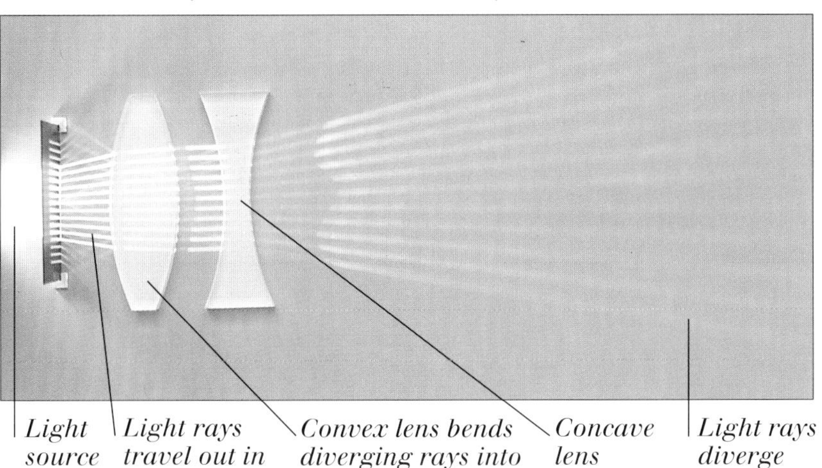

CONVEX LENS (BENDS LIGHT INWARDS)

Light rays converge

First convex lens produces parallel beams

Convex lens

Focal length

Light focused to a point

CONVEX MIRROR (REFLECTS LIGHT OUTWARDS)

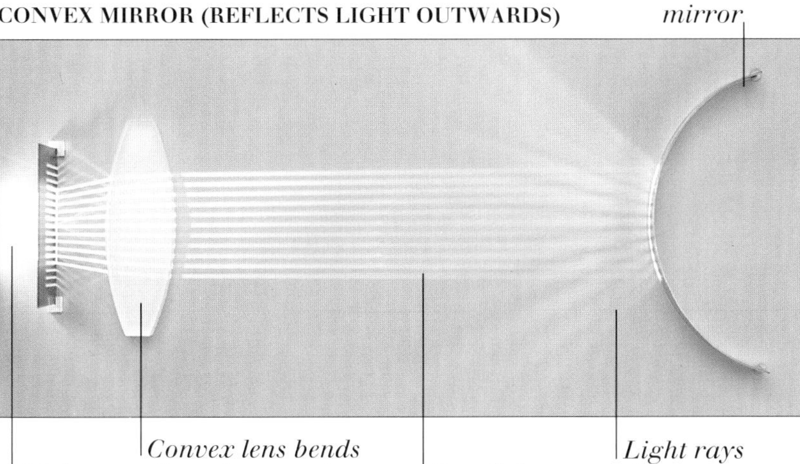

CONCAVE MIRROR (REFLECTS LIGHT INWARDS)

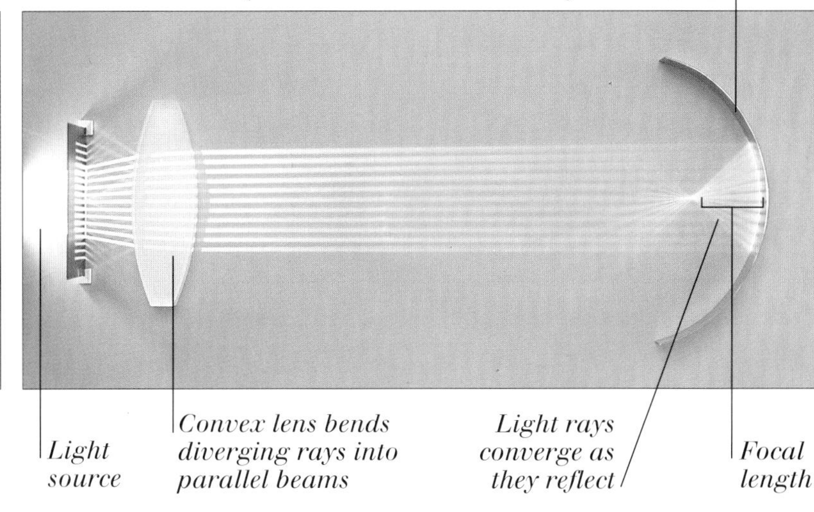

LENSES

Concave lenses make objects appear smaller, and allow a larger field of vision. Objects lying within the focal length of a convex lens appear larger.

CONCAVE LENS

Regular squares

A concave lens is often fitted to the rear window of a vehicle to improve a driver's field of vision

Squares appear smaller through lens

CONVEX LENS

A convex lens can be used as a magnifying glass

Regular squares

Squares appear magnified through lens

IMAGE FORMATION

Because they focus light, convex lenses can be used to project images onto a screen. The screen must be placed at a point where the rays focus in order for a clear image to be produced. Only objects that lie within a range of distances from the lens, called the **depth of field**, will be in focus at any one time.

IMAGE INVERTS

Optical axis

Ray 1 starts parallel to optical axis

Ray 3 goes through the focal point in front of the lens

Convex lens

Ray 2 goes through centre of lens so is undeviated

Ray 1 is bent and goes through focal point of lens

Ray 3 is bent parallel to the optical axis

The rays focus on the opposite side of the optical axis so the image is inverted

PROJECTED IMAGE

Convex lens

Black arrows drawn on tracing paper

Screen

Focused image on screen

Image is inverted vertically and horizontally

Optical instruments

THE HUMAN EYE CONTAINS a **lens** that produces an **image** by focusing the light that passes through it. But the eye does not record images, or allow us to see objects that are very small, or very far away. To achieve these tasks, we need to use optical instruments. A camera, for example, records an image on light-sensitive film. To see objects that are very small or very far away, we need to produce a magnified image, which the eye can then observe. By using a compound microscope, light from a very small object can be made to produce a magnified image. A telescope produces a magnified image in a similar way to a microscope, using lenses to focus light. There are limits to the use of optical instruments. Even the most precise lenses suffer from **chromatic aberration** (see opposite) – a problem that can be solved by pairs of lenses known as **achromatic doublets**.

CAMERA

The cutaway view below shows the main features of a single lens reflex (SLR) camera. The light is focused onto film at the back of the camera by a lens or a combination of lenses.

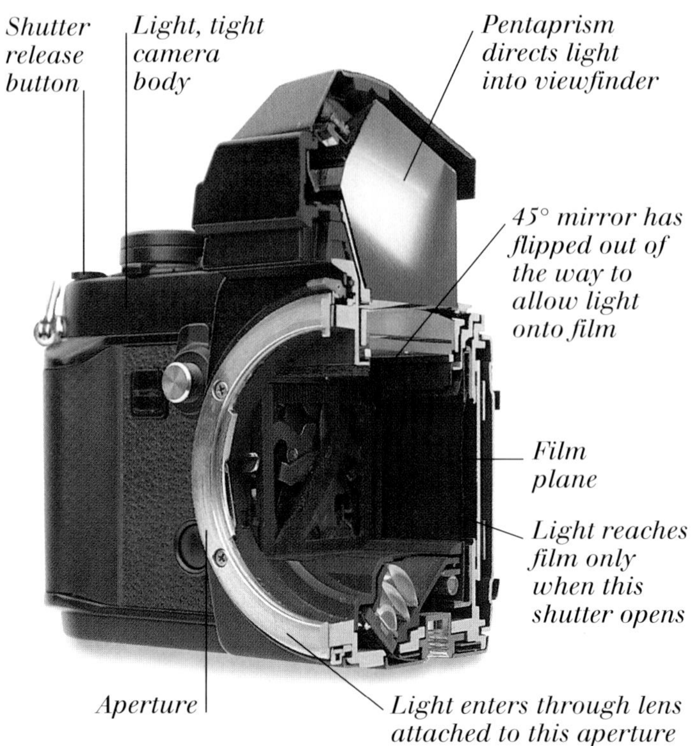

MICROSCOPES

A basic optical microscope consists of two lenses, the objective and the eyepiece. Light is focused by the objective, which has a very small **focal length**. The light passes up inside the body tube, and is focused by the eyepiece into the eye.

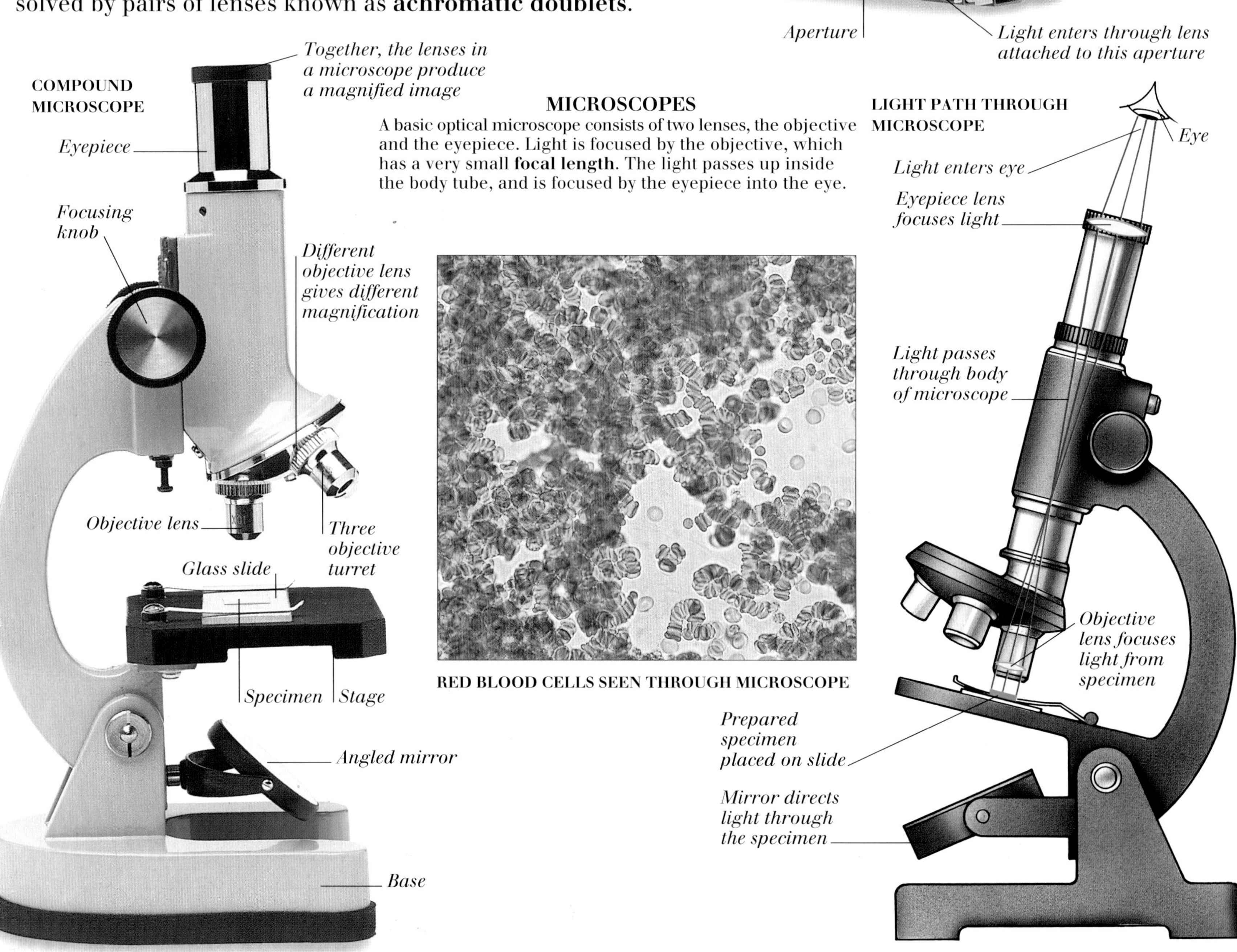

RED BLOOD CELLS SEEN THROUGH MICROSCOPE

TELESCOPES

At the front of a **refracting** telescope is an objective lens that collects light and focuses it, producing an image in the telescope's tube. The eyepiece greatly magnifies this image. A **reflecting** telescope uses a mirror instead of an objective lens.

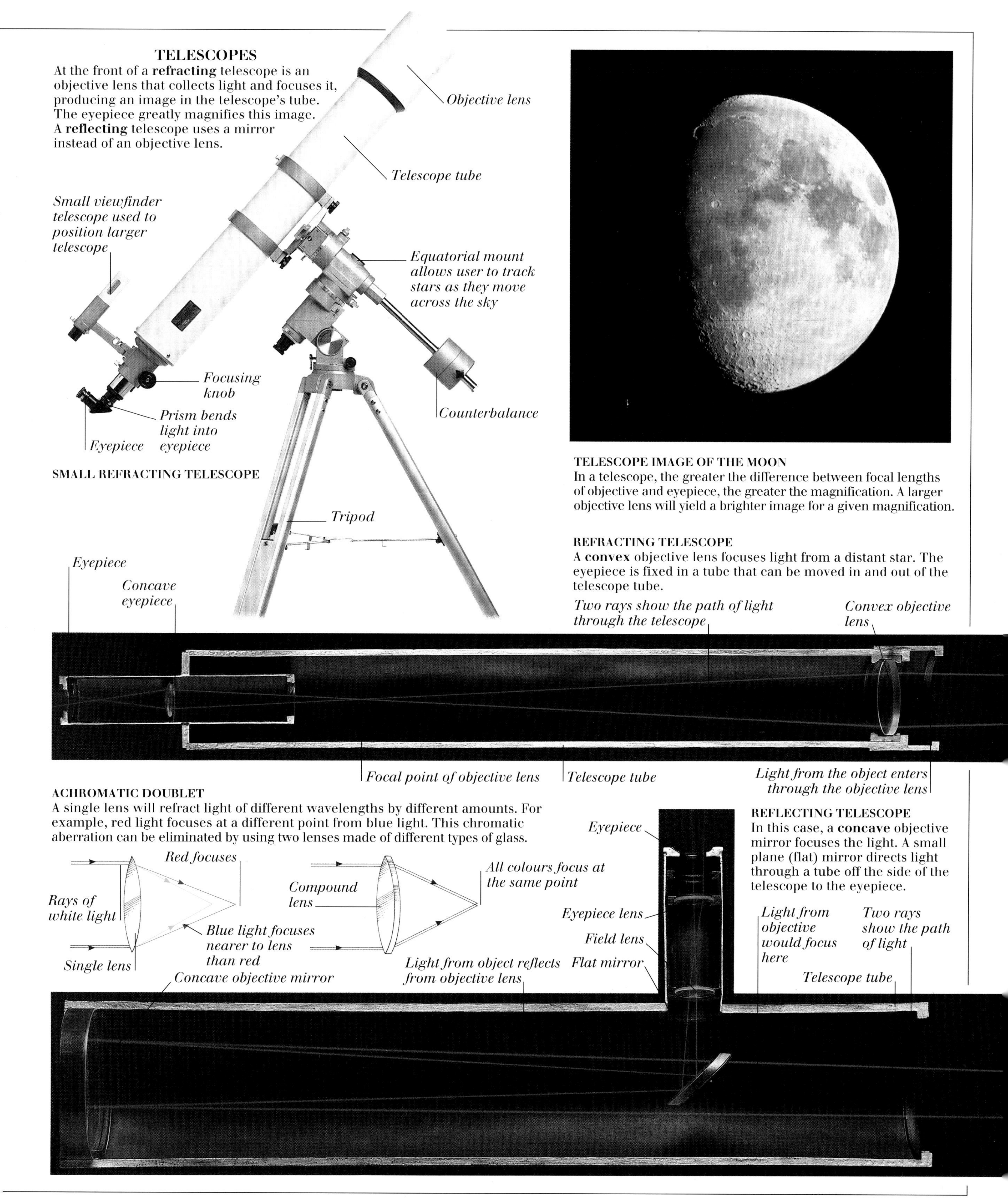

SMALL REFRACTING TELESCOPE

TELESCOPE IMAGE OF THE MOON

In a telescope, the greater the difference between focal lengths of objective and eyepiece, the greater the magnification. A larger objective lens will yield a brighter image for a given magnification.

REFRACTING TELESCOPE

A **convex** objective lens focuses light from a distant star. The eyepiece is fixed in a tube that can be moved in and out of the telescope tube.

ACHROMATIC DOUBLET

A single lens will refract light of different wavelengths by different amounts. For example, red light focuses at a different point from blue light. This chromatic aberration can be eliminated by using two lenses made of different types of glass.

REFLECTING TELESCOPE

In this case, a **concave** objective mirror focuses the light. A small plane (flat) mirror directs light through a tube off the side of the telescope to the eyepiece.

Wave behaviour

ALL TYPES OF WAVE CAN COMBINE OR INTERFERE. If two waves are in step, so that the peaks coincide, the **interference** results in a wave that will be larger than the original one (**constructive interference**). If the waves are out of step, the peak of one wave will cancel out the trough of another (**destructive interference**). Where the waves are equal in size, they can cancel out entirely. As waves pass around objects or through small openings, they can be **diffracted**, or bent. Diffraction and interference can be observed in water waves, using a ripple tank. The colours seen in soap bubbles are the result of some colours being removed from the white light spectrum by destructive interference. Light reflected off the front and back surfaces of the film interferes. One source of very pure light waves is a laser. The light produced by a laser is coherent. This means that all of the waves are in step and of exactly the same wavelength. The light is produced by a process called stimulated emission. To understand this process, light must be thought of in terms of particles called **photons** (see p.38), as well as waves.

PRINCIPLE OF SUPERPOSITION

When two waves meet, they add up or interfere, combining their separate values. This is called the **Principle of Superposition**, and is common to all types of wave.

CONSTRUCTIVE INTERFERENCE

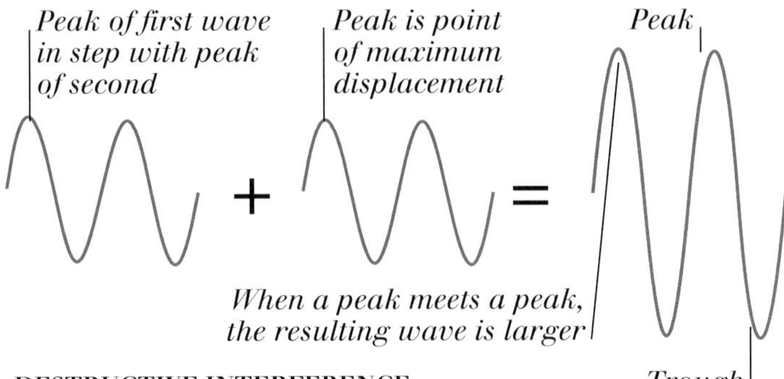

DESTRUCTIVE INTERFERENCE

DIFFRACTION AND INTERFERENCE

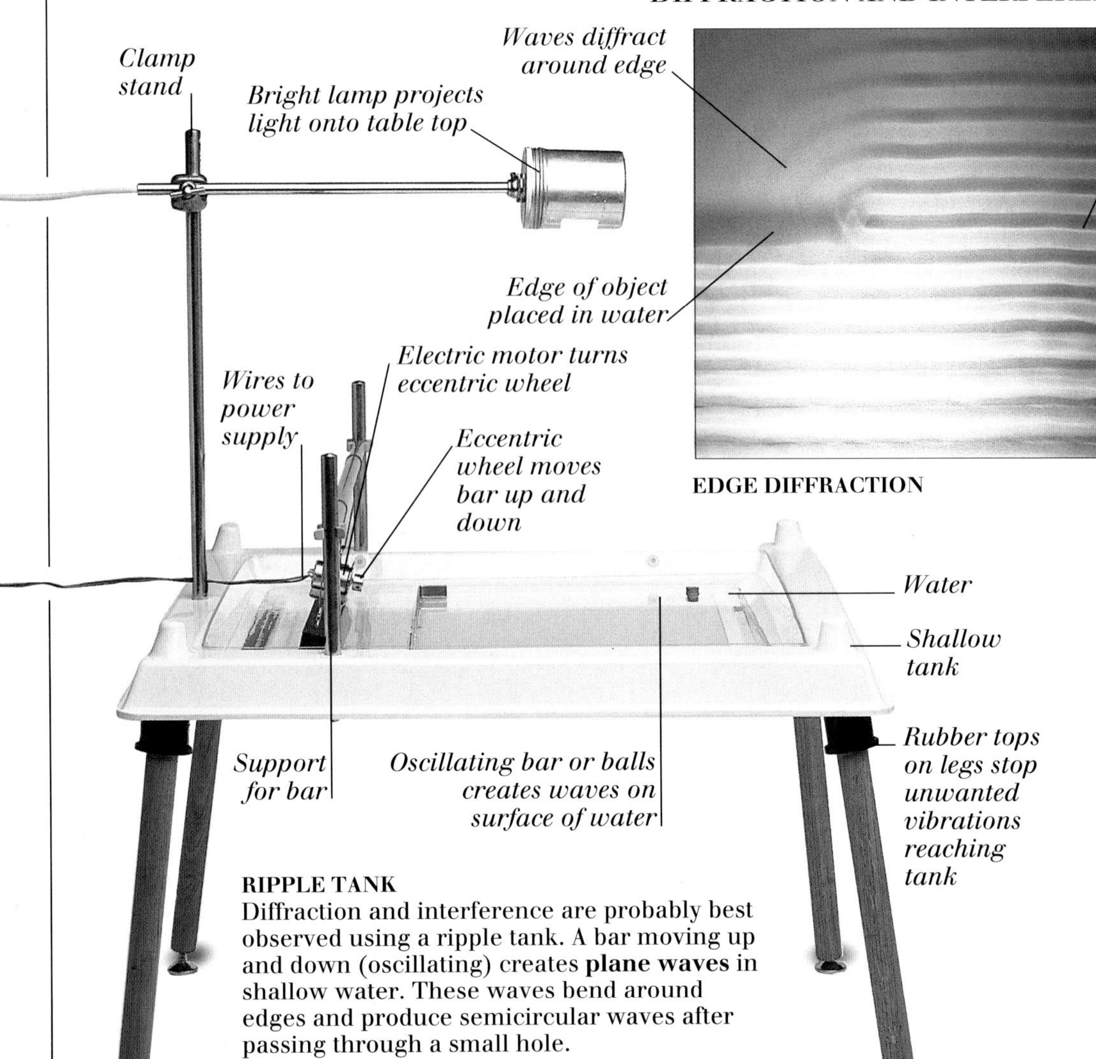

RIPPLE TANK
Diffraction and interference are probably best observed using a ripple tank. A bar moving up and down (oscillating) creates **plane waves** in shallow water. These waves bend around edges and produce semicircular waves after passing through a small hole.

EDGE DIFFRACTION

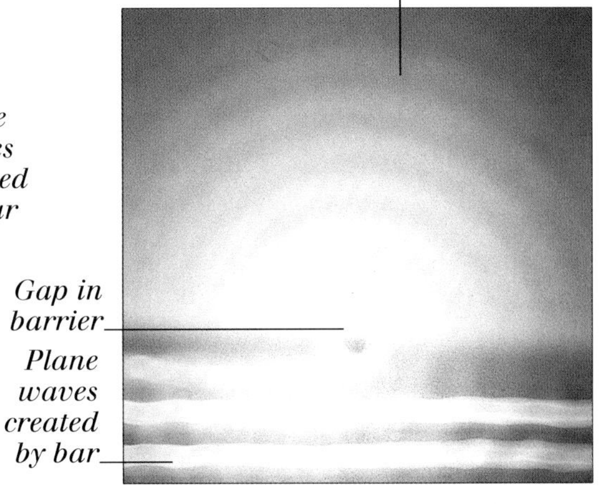

DIFFRACTION THROUGH SMALL HOLE

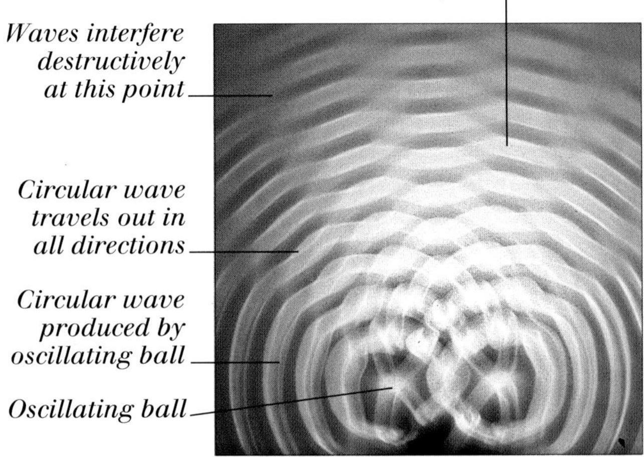

INTERFERENCE

THIN FILM INTERFERENCE

White light reflects off the front and back surfaces of a soap film. The two reflected beams of light interfere. Some wavelengths, and therefore some colours, will be lost from the white light by destructive interference. Which colours are lost depends on the thickness of the film.

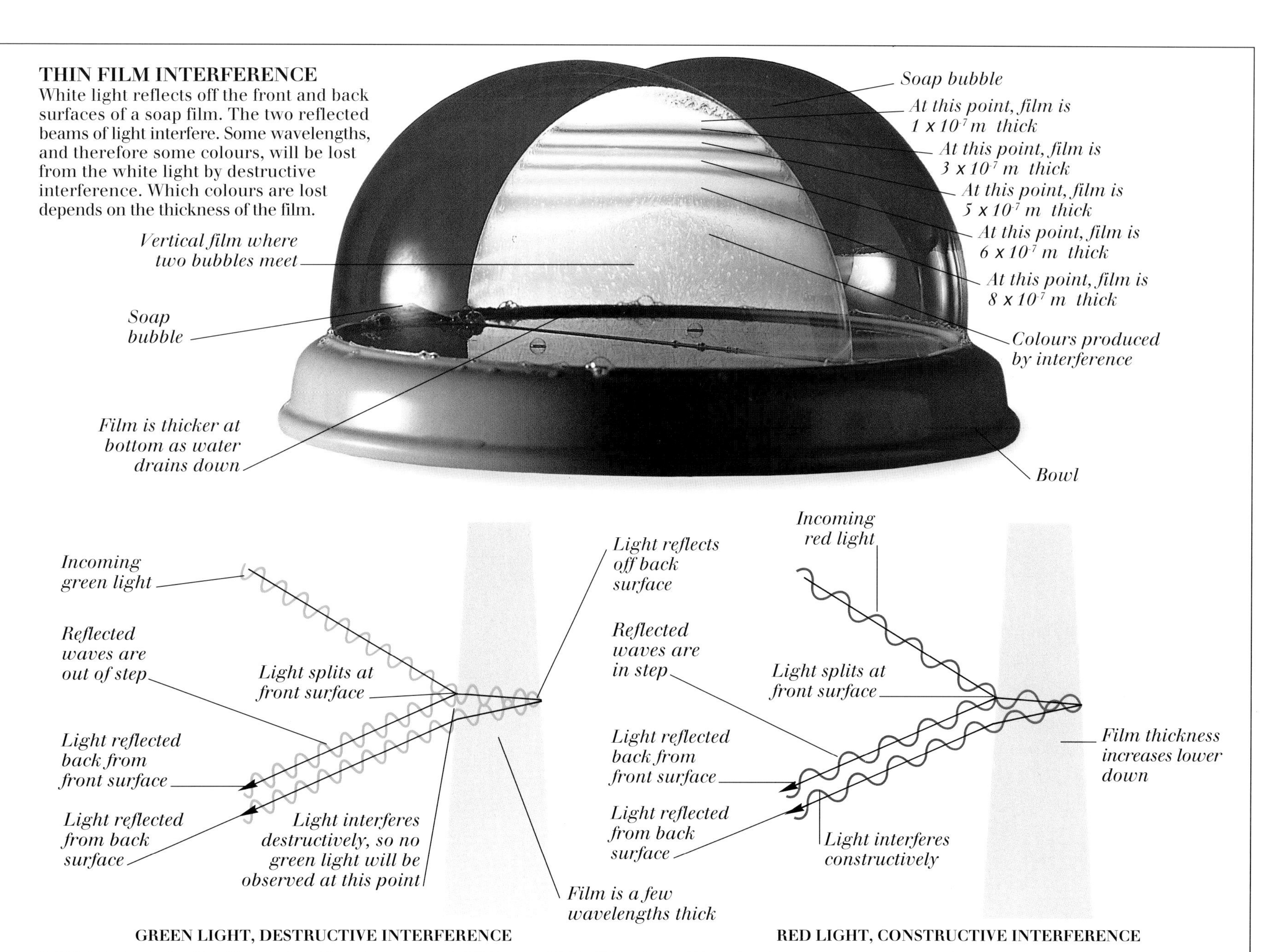

GREEN LIGHT, DESTRUCTIVE INTERFERENCE

RED LIGHT, CONSTRUCTIVE INTERFERENCE

LASERS

STIMULATED EMISSION

Light behaves as waves and particles (see p. 38). Laser light exhibits all of the behaviour common to waves, including **interference** and **diffraction**. But to understand the operation of a laser, light must be thought of as being composed of particles called **photons**. Each photon is emitted as the result of the stimulation of an excited **electron** by another photon within the laser.

LASER SURGERY

Lasers have found many applications in medicine. Here, a laser is being used to destroy a cateract in a patient's eye.

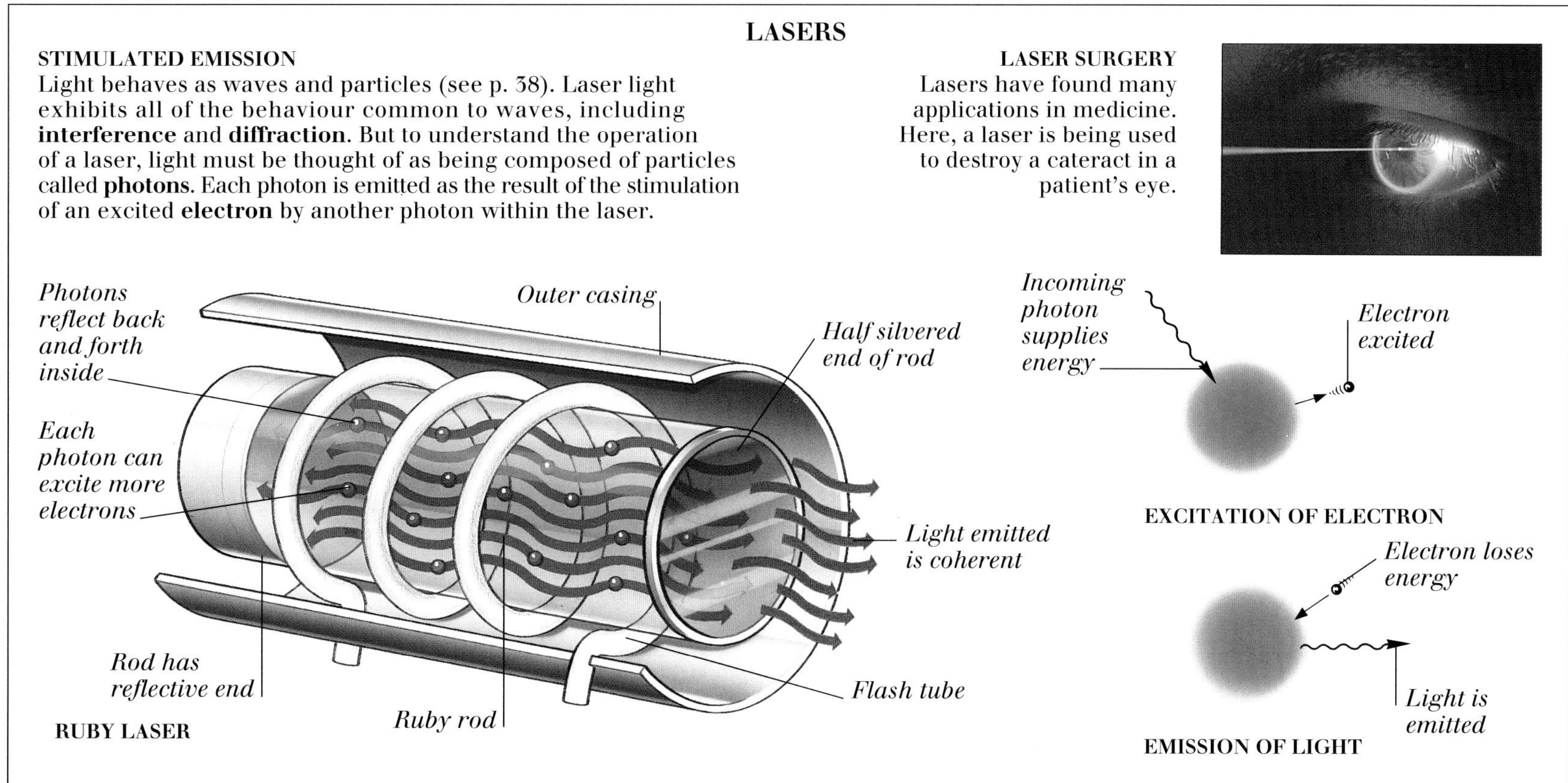

RUBY LASER

EXCITATION OF ELECTRON

EMISSION OF LIGHT

Atoms and electrons

ALL ORDINARY MATTER consists of tiny particles called **atoms**. There are 92 naturally occurring types of atom. Each consists of a central, positively charged **nucleus** (see pp. 50-51), surrounded by negatively charged **electrons**. An **element** is a substance made up of one type of atom only. Atoms of different elements have different numbers of electrons. For example, atoms of the element fluorine have nine electrons. Electrons in the atom do not follow definite paths, as planets do, orbiting the Sun. Instead, they are said to be found in regions called **orbitals**. Electrons in orbitals close to the nucleus have less energy than those farther away and are said to be in the first electron **shell**. Electrons in the second shell have greater energy. Whenever an **excited** electron releases its energy by falling to a lower shell, the energy is radiated as light. This is called **luminescence**. Electrons can be separated from atoms in many ways. In a **cathode ray tube**, a strong electric field tears electrons away from their atoms. Free electrons in the tube are affected by **electric** and **magnetic fields**. Cathode ray tubes are used in television, where a beam of free electrons forms the picture on the screen.

ANATOMY OF A FLUORINE ATOM

A fluorine atom has nine electrons around its nucleus. There are two electrons in the first shell, in an s-orbital (1s). The remaining seven electrons are found in the second shell, two in an s-orbital (2s) and five in p-orbitals (2p).

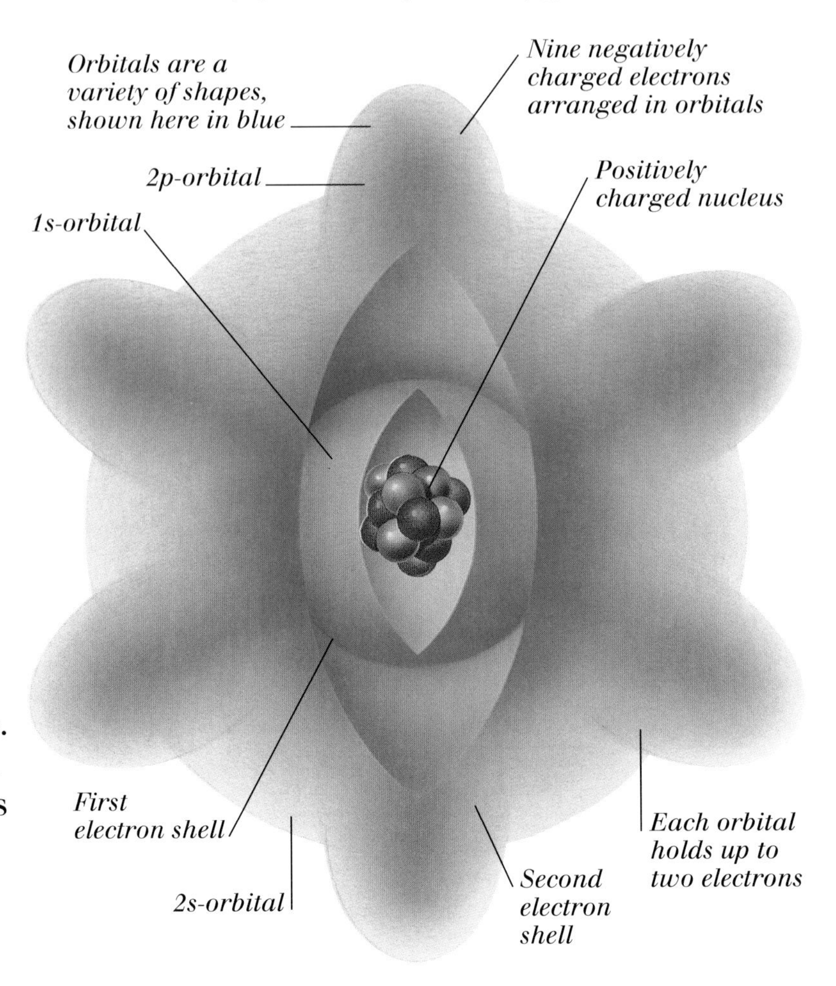

FLUORESCENCE

WASHING POWDER

How does washing powder make clothes appear so bright? Washing powder exhibits a form of luminescence (see above) called **fluorescence**. Electrons around atoms in the powder are excited into high-energy shells by incoming energy, in this case invisible ultraviolet light found in daylight. The electrons fall back to their original shell immediately and re-emit the energy as visible light. The colour of the light depends upon the difference in energy between the higher and lower shells. This extra light emitted by the powder in the clothes makes them appear bright.

WASHING POWDER IN WHITE LIGHT

WILLEMITE IN WHITE LIGHT

SODALITE IN WHITE LIGHT

WASHING POWDER IN ULTRAVIOLET LIGHT

WILLEMITE IN ULTRAVIOLET LIGHT

SODALITE IN ULTRAVIOLET LIGHT

ELECTRON BEAMS

CATHODE RAY TUBE
Inside a cathode ray tube, an **electric current** heats a small filament. The heat generated gives electrons extra energy, moving them farther from their nuclei. A strong electric field then completely removes electrons from their atoms. The free electrons are attracted to the positive **anode**, and pass through it as a cathode ray.

DEFLECTING THE ELECTRONS
Because electrons have electric charge, forces can be applied to them by electric and magnetic fields in the cathode ray tube. The direction of the force depends upon the direction and type of the field.

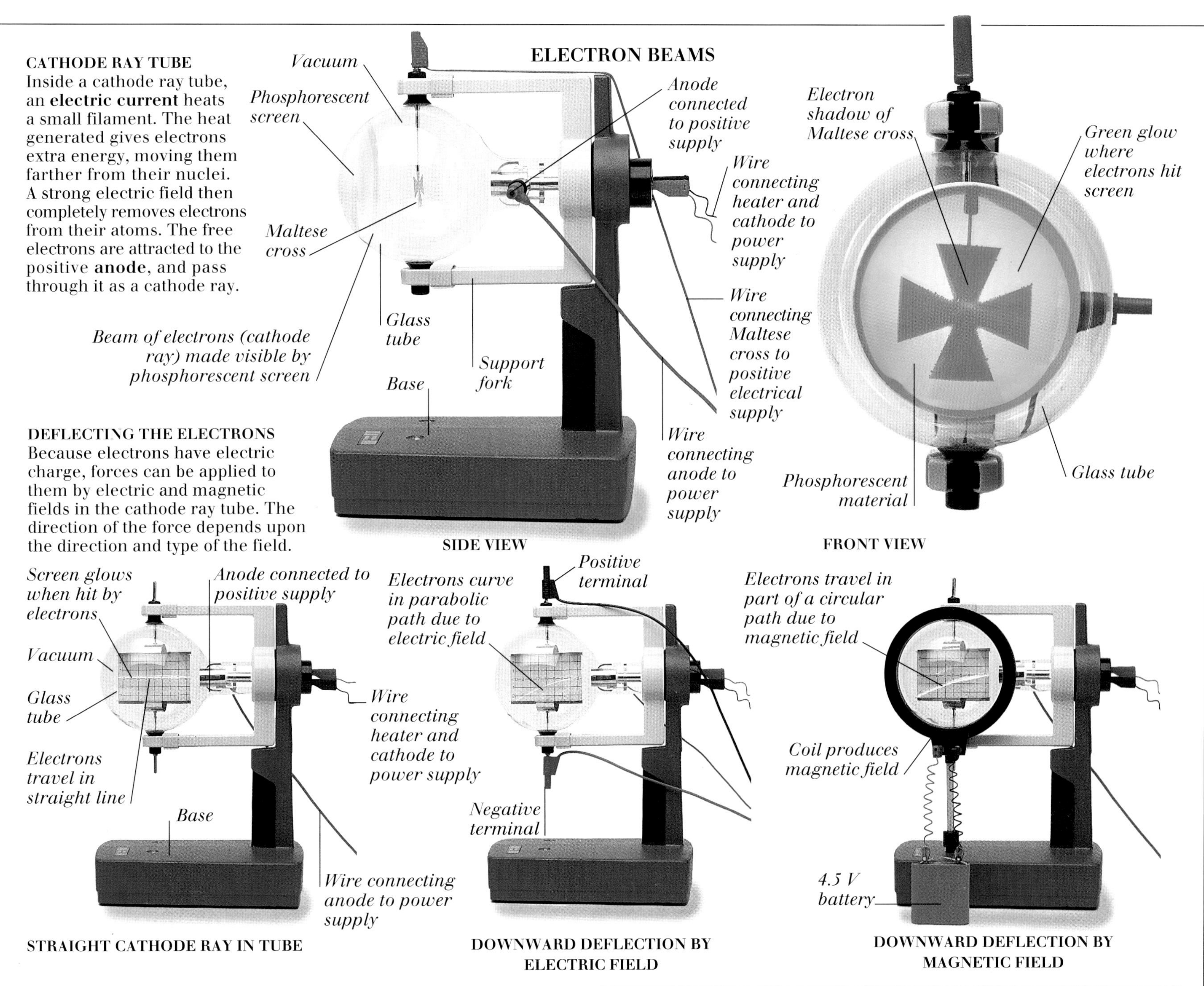

SIDE VIEW

FRONT VIEW

STRAIGHT CATHODE RAY IN TUBE

DOWNWARD DEFLECTION BY ELECTRIC FIELD

DOWNWARD DEFLECTION BY MAGNETIC FIELD

HOW A TELEVISION WORKS

DEFLECTED ELECTRON BEAMS
At the heart of most televisions is a cathode ray tube. Electron beams are produced at the back of the tube. Coils of wire around the tube create magnetic fields, which deflect the electron beams to different parts of the screen. The screen itself is coated with phosphorescent materials called **phosphors**.

PHOSPHORESCENCE
When the cathode rays hit the special coating on a television screen, the screen glows because it is phosphorescent. **Phosphorescence** is a form of luminescence where the incoming energy is not re-emitted immediately but is stored and re-emitted over a period of time. This means that as the cathode ray quickly scans the picture, the phosphor glows for long enough for a whole picture to form.

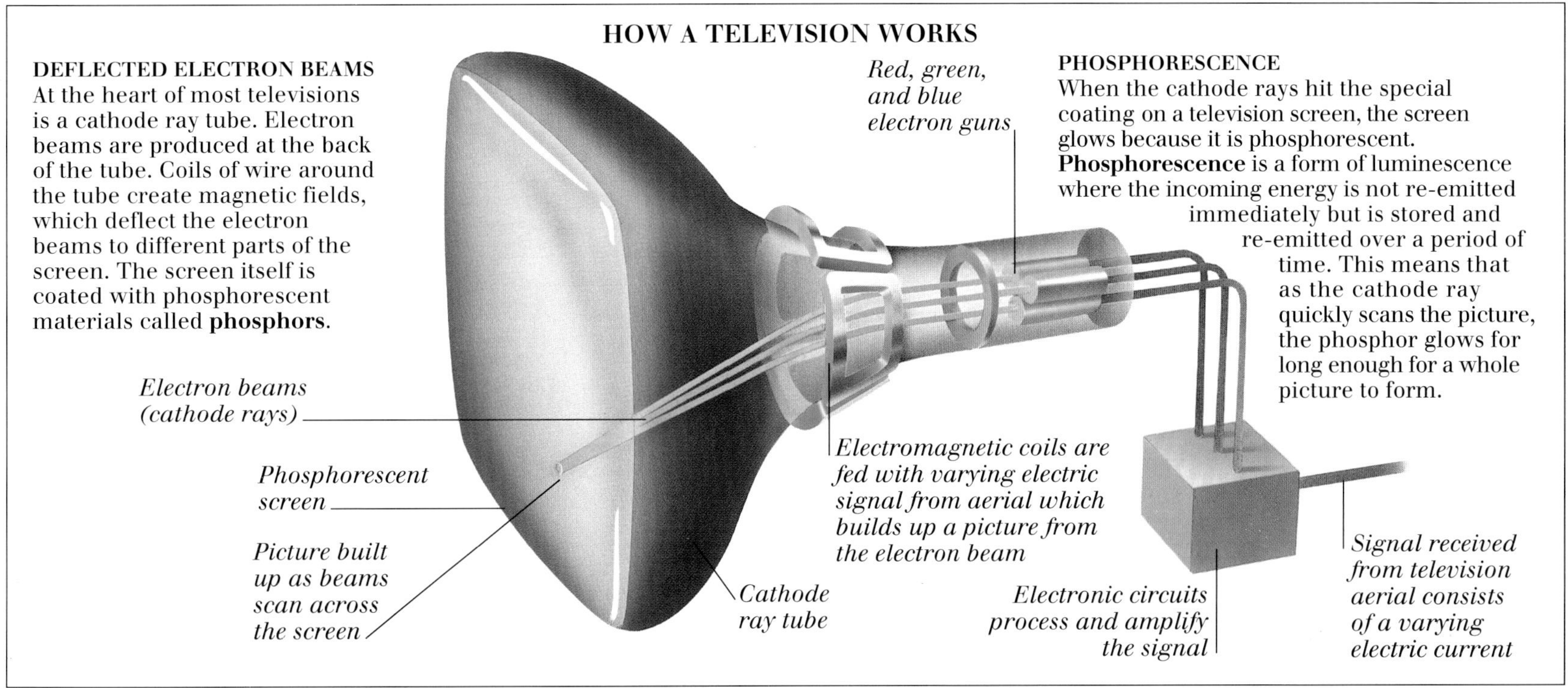

Nuclear physics

AT THE CENTRE OF EVERY ATOM LIES a positively charged **nucleus**. It consists of **protons** and **neutrons**. The number of protons in the nucleus is called the **atomic number**. Because they all have the same **electric charge**, protons repel each other. The nucleus holds together despite this repulsion because of the **strong nuclear force** (see pp. 52-53). The balance between the repulsive force and the strong nuclear force determines whether a nucleus is stable or unstable. On the whole, small nuclei are more stable than larger ones, because the strong nuclear force works best over small distances. An unstable, larger nucleus can break up or decay in two main ways, **alpha decay** and **beta decay**. These produce **alpha** and **beta particles**. In each type of decay, the atomic number of the new nucleus is different from the original nucleus, because the number of protons present alters. Nuclei can also completely split into two smaller fragments, in a process called **fission**. In another **nuclear reaction** called **fusion**, small nuclei join together. Both of these reactions can release huge amounts of energy. Fusion provides most of the Sun's energy, while fission can be used in power stations to produce electricity.

FLUORINE-19 NUCLEUS
The number of protons in a nucleus defines what **element** the atom is. For example, all fluorine atoms have nine protons. Fluorine has an atomic number of 9. The number of neutrons can vary. Fluorine-19 has ten neutrons, while fluorine-18 has nine.

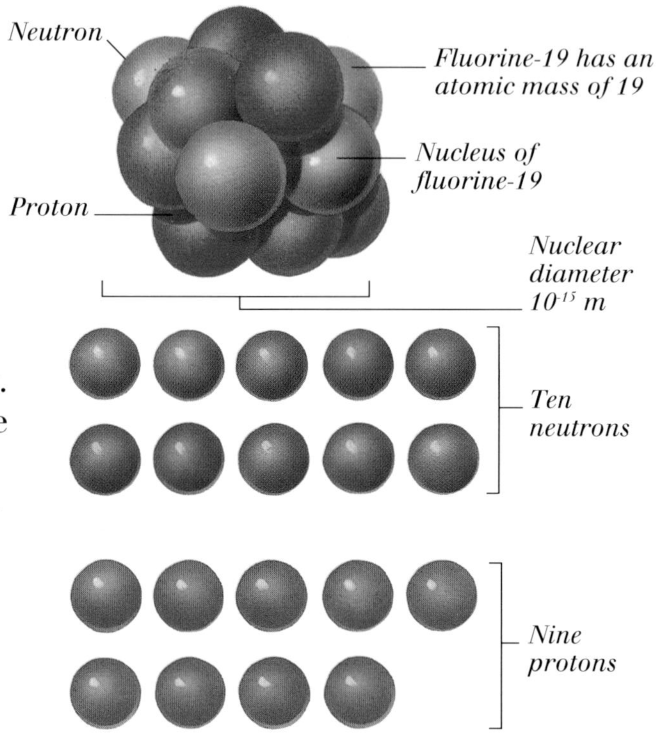

RADIOACTIVITY

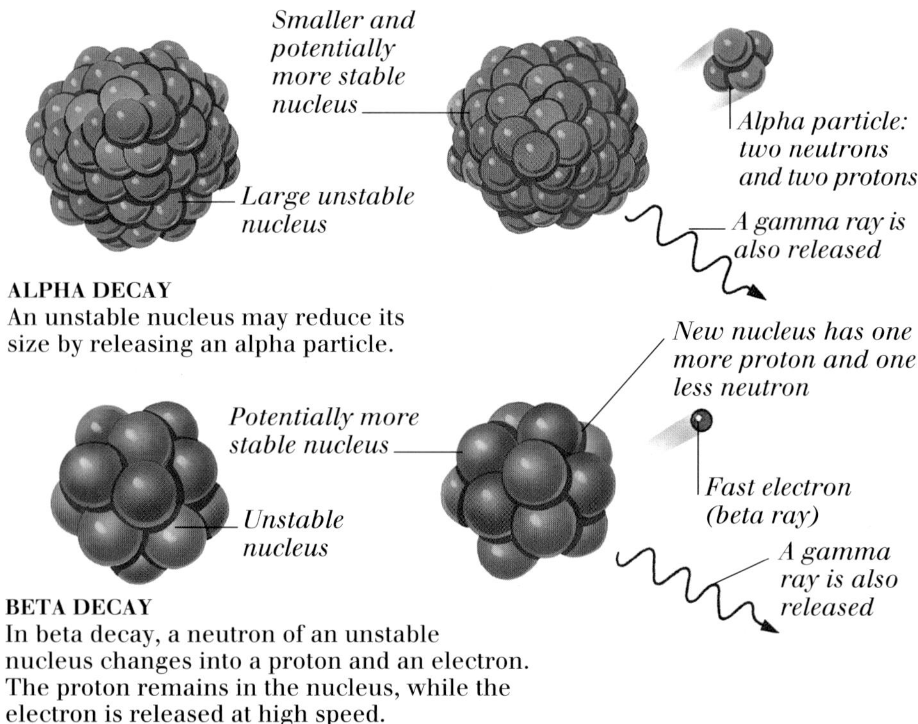

ALPHA DECAY
An unstable nucleus may reduce its size by releasing an alpha particle.

BETA DECAY
In beta decay, a neutron of an unstable nucleus changes into a proton and an electron. The proton remains in the nucleus, while the electron is released at high speed.

ANALYSING RADIOACTIVITY
Because of their electric charges, a strong magnetic field will deflect alpha and beta rays into curved paths. Cloud chambers are used to show these paths, as in the illustration below.

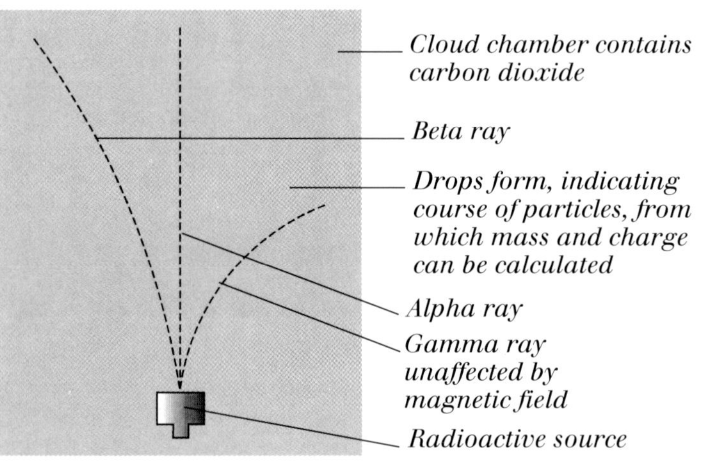

GEIGER-MULLER TUBE
As they pass through the air, alpha and beta rays hit atoms, separating electrons and creating **ions**, which can be detected inside a Geiger-Müller tube.

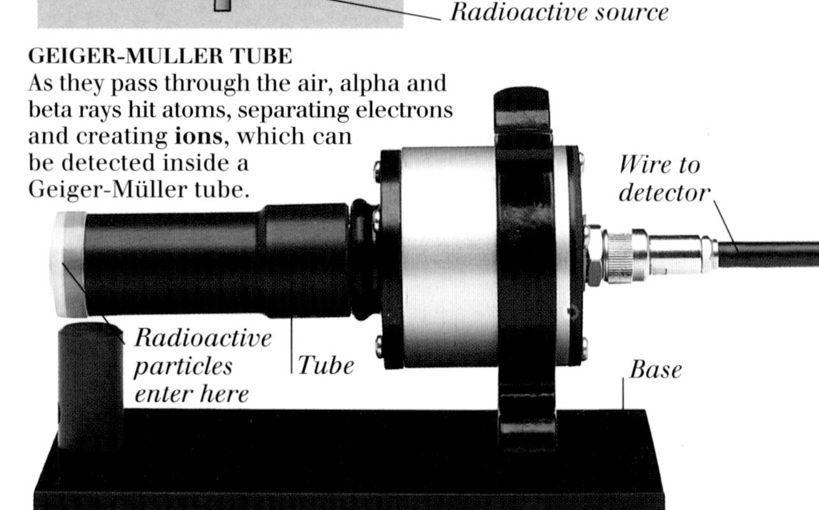

COSMIC RAYS
The Earth is constantly bombarded by particles from space. Most of these are protons, from atoms of the most abundant element, hydrogen. Occasionally, the protons collide with atoms in the air, producing showers of secondary particles called cosmic rays.

Tracks left by cosmic rays in a bubble chamber

NUCLEAR FISSION

A neutron hitting a large, unstable nucleus may split or fission into two smaller, more stable fragments, releasing large amounts of energy. Often, more free neutrons are produced by this fission, and these neutrons can cause other nuclei to split. The process may continue, involving many nuclei in a **chain reaction**.

NUCLEAR FUSION

Just as large nuclei can split, so some small nuclei can join together, or fuse. Like fission, fusion can release energy. One of the highest energy fusion reactions involves nuclei of hydrogen, which collide at great speed, forming a nucleus of helium.

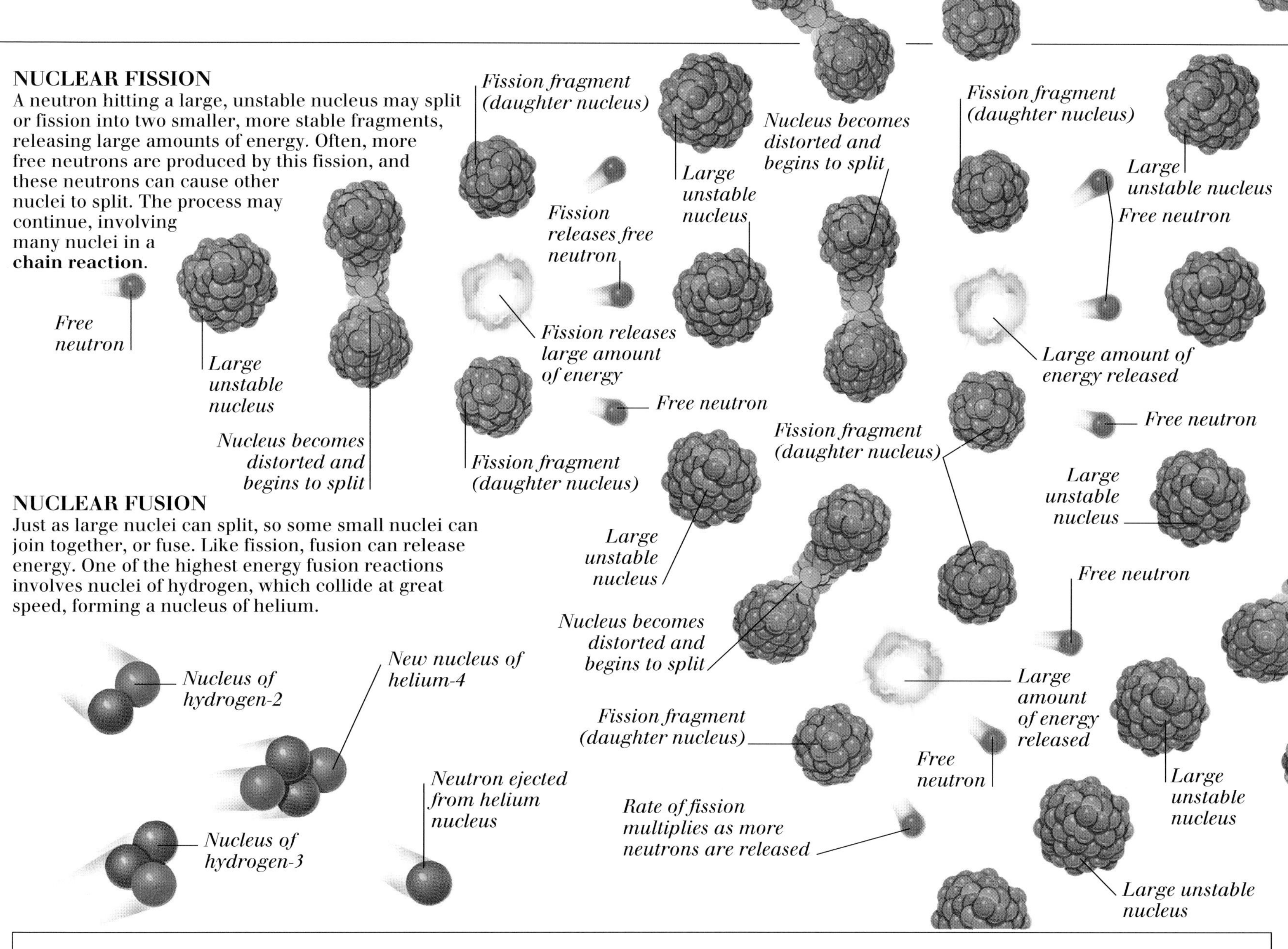

NUCLEAR POWER

NUCLEAR POWER STATION

A nuclear chain reaction releases huge amounts of heat. This heat can be used to generate electricity (see pp. 36-37), in a nuclear power station. The reactions occur in the nuclear reactor, and the heat produced is used to make steam.

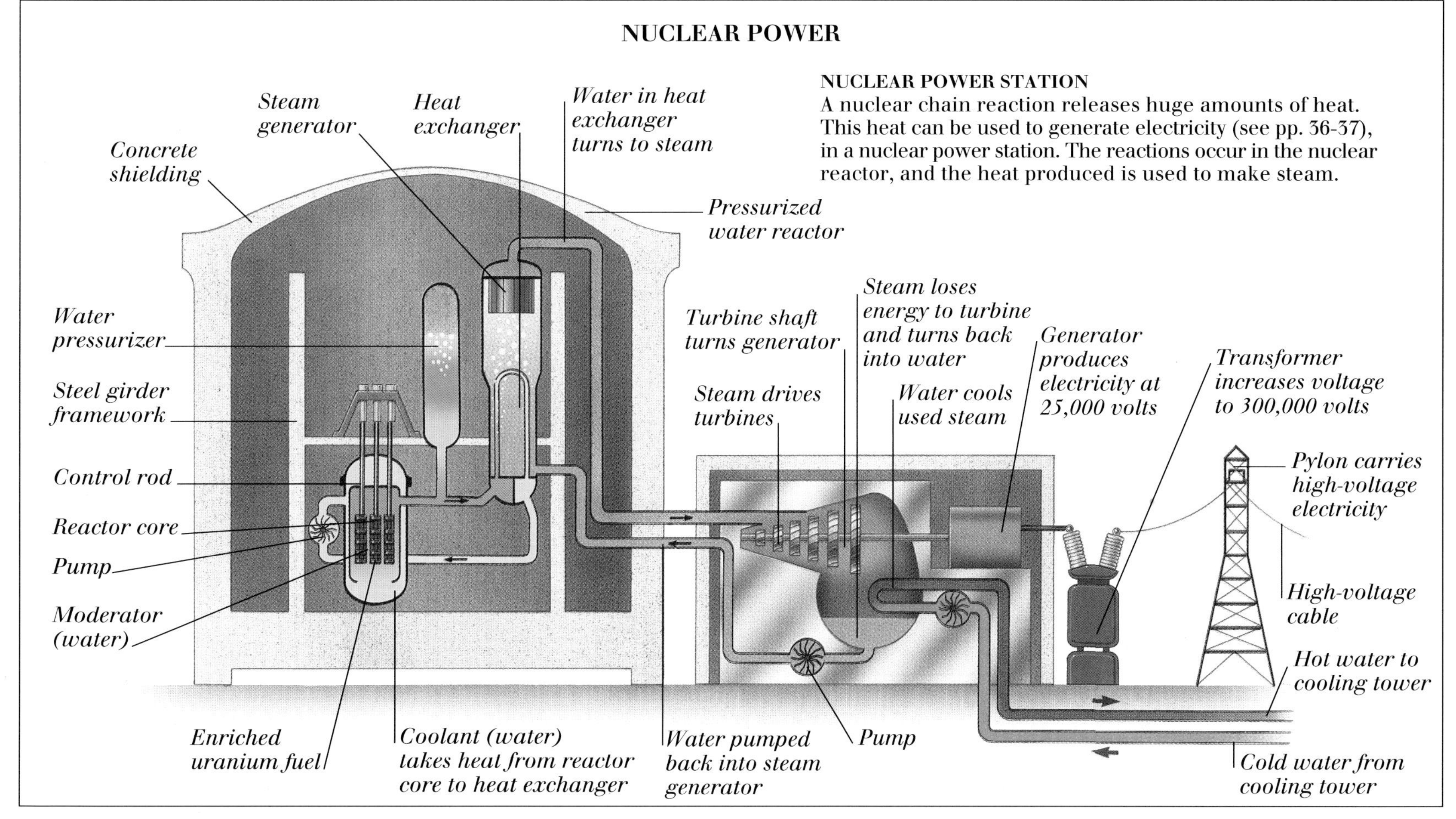

Particle physics

PARTICLE PHYSICS ATTEMPTS TO EXPLAIN matter and **force** in terms of tiny particles. The **atom**, once thought to be the smallest particle, is actually made of **protons, neutrons,** and **electrons**. But the proton and the neutron are themselves made up of smaller particles, known as **quarks**. There are four types of force acting between matter, namely **gravitational force**, the **electromagnetic force**, the **strong nuclear force,** and the **weak interaction**. According to current theory, each of these forces is explained by the exchange of particles called **gauge bosons** between the particles of matter. For example, the **nucleus** holds together as a result of the exchange of particles called **mesons** (a type of gauge boson) between the protons and neutrons present. These exchanges can be visualized in Feynman diagrams, which show the particles involved in each type of force. The most important tools of particle physics are particle accelerators, which create and destroy particles in high-energy collisions. Analysis of these collisions helps to prove or disprove the latest theories about the structure of matter and the origin of forces. One of the current aims of large particle accelerators, such as the Large Hadron Collider at **CERN** (see opposite), is to prove the existence of a particle called the **Higgs boson**. It may be responsible for giving all matter mass.

HADRONS

Protons, neutrons, and mesons are examples of hadrons. A **hadron** is a particle consisting of quarks. There are six types of quark, including the "up" and "down" quarks. The quarks of hadrons are held together by **gluons**.

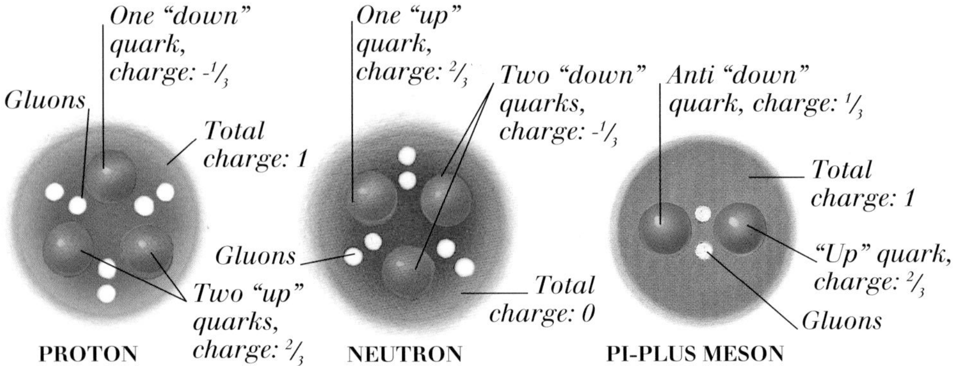

FEYNMAN DIAGRAMS

The diagrams below show which gauge bosons are exchanged to transfer each of the four types of force. The horizontal lines represent the gauge boson, whereas the diagonal lines and the circles represent the two interacting particles.

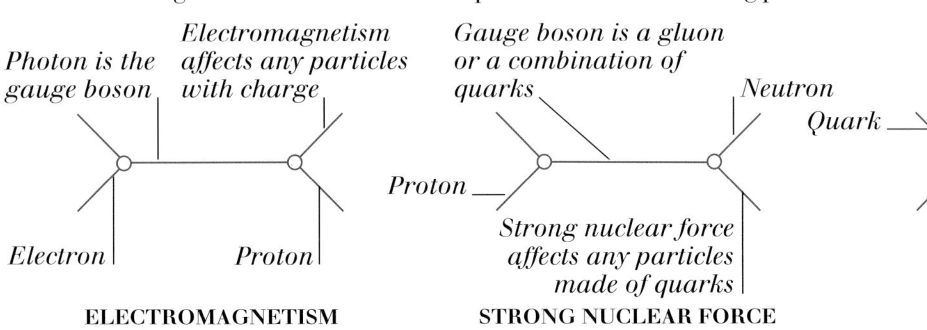

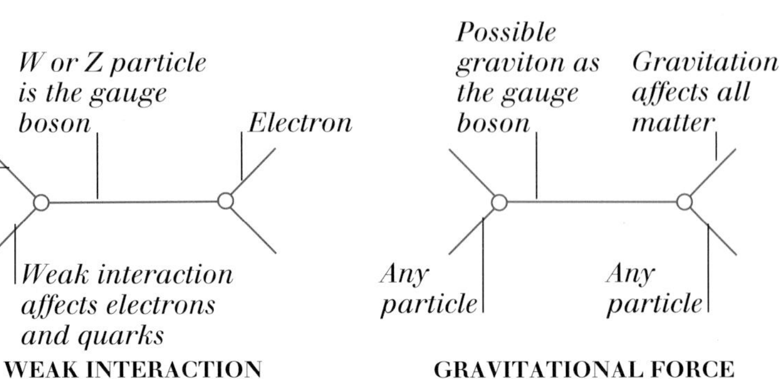

PARTICLE COLLISIONS

The images below show the results of collisions between particles in particle accelerators. Particles of opposite charge curve in different directions in the strong **magnetic field** of the detector.

ANNIHILATION
When a particle and an antiparticle meet, they destroy each other and become energy. This energy in turn becomes new particles.

PROTON-PHOTON COLLISION
This collision between a **photon** and a proton took place in a type of detector called a bubble chamber. The colours in this photograph have been added for clarity.

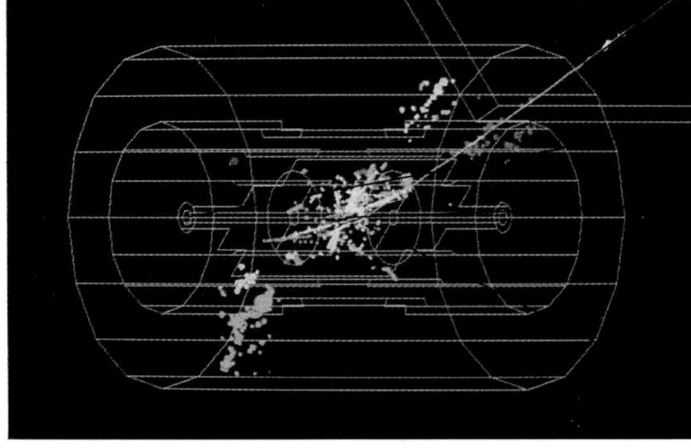

ELECTRON-POSITRON COLLISION
Here, an electron collides with its **antiparticle**, a **positron**. The detector is linked to a computer, which produces this picture of the collision.

THE LARGE HADRON COLLIDER

MAP OF THE SITE

The Large Hadron Collider (LHC), at CERN near Geneva, will be a huge particle accelerator, in a tunnel about 100 metres below ground. The tunnel will be a ring 27 kilometres long, which is already used for another particle accelerator, the Large Electron Positron (LEP) collider. Two beams of protons will move around in tubes at very high speed, and will be made to collide in detectors, such as the CMS (see below).

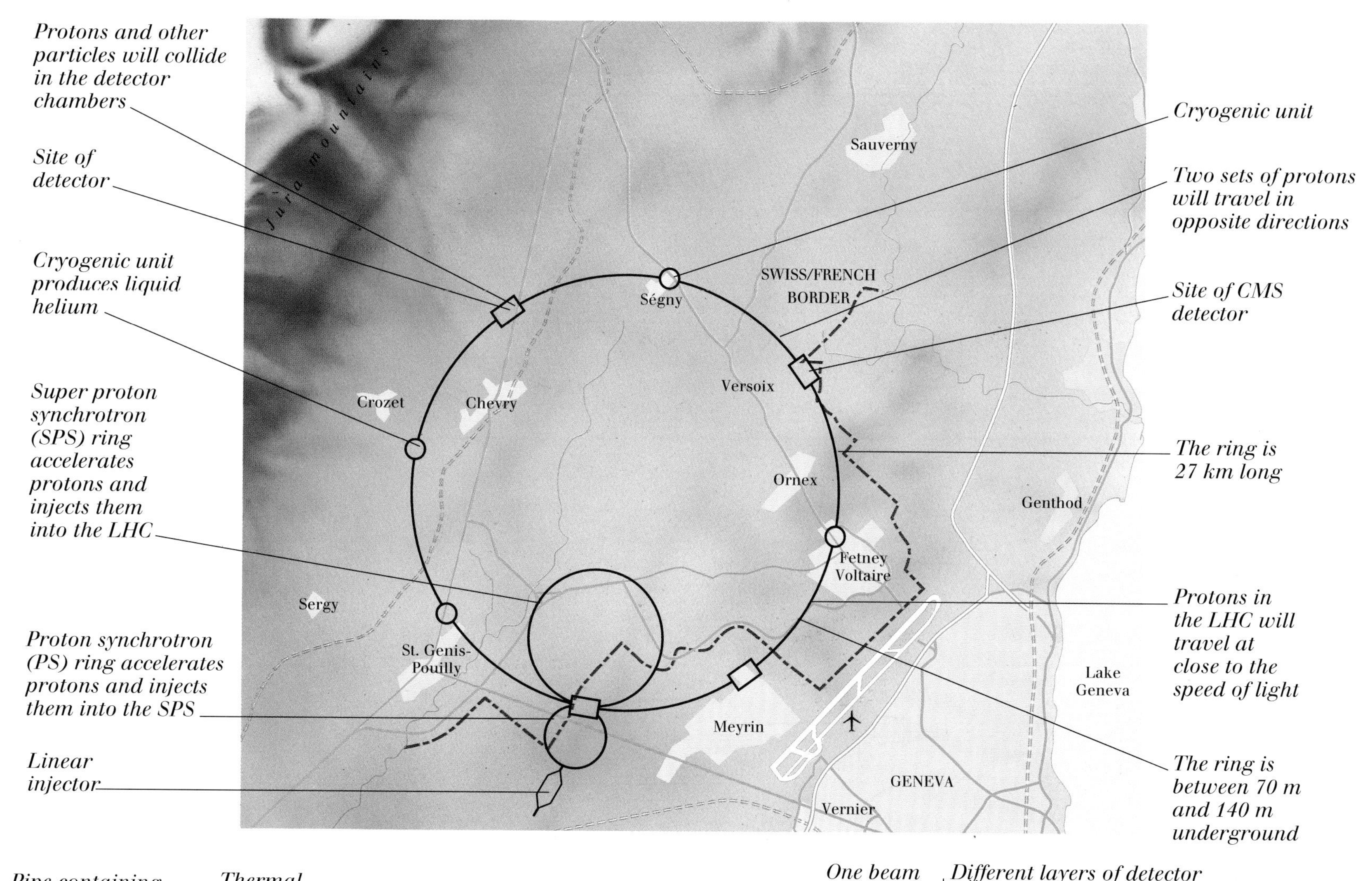

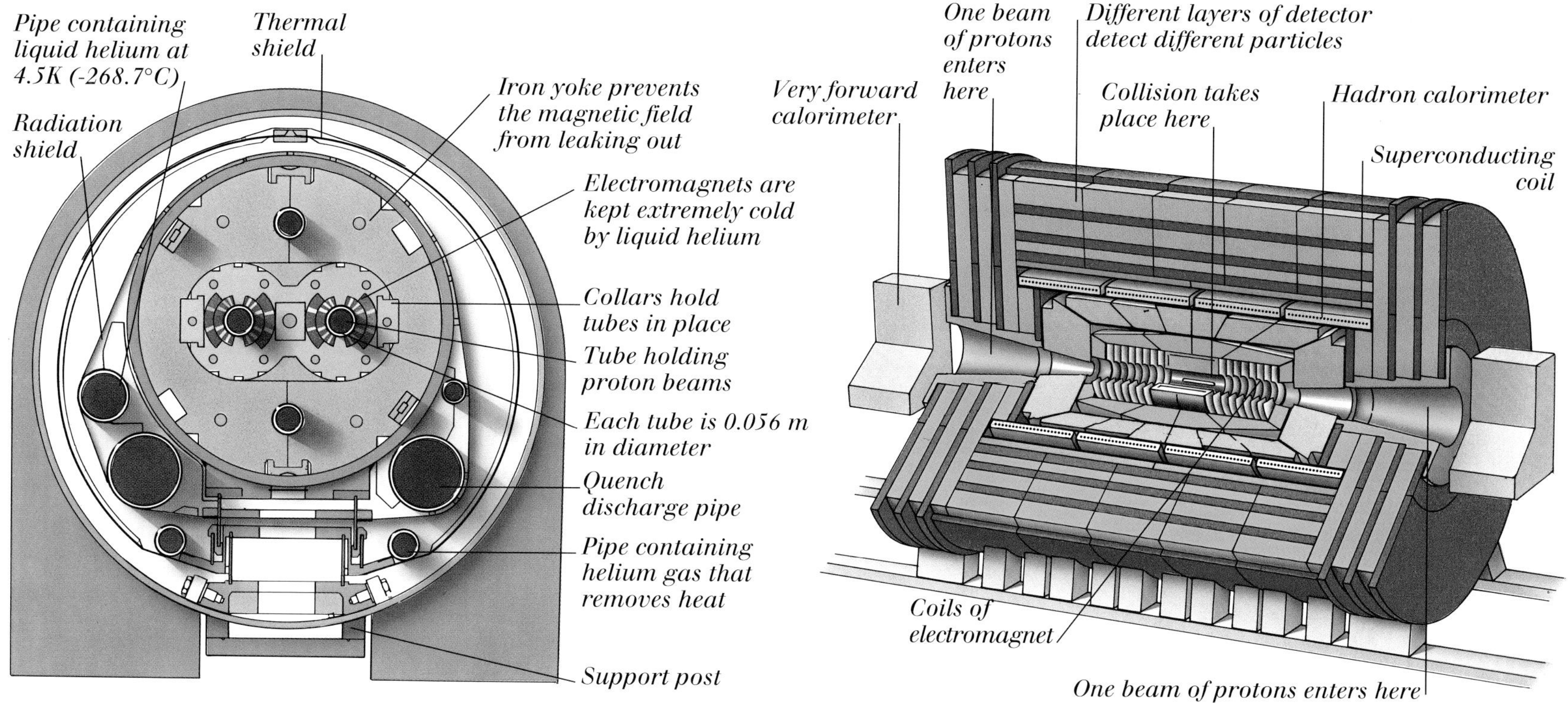

THE ACCELERATOR

In the main experiment of the LHC, protons injected into the ring will be accelerated to nearly the speed of light, travelling in opposite directions in two tubes. **Centripetal force** provided by powerful **electromagnets** keeps the protons moving in a circle.

THE COMPACT SOLENOIDAL (CMS) DETECTOR

Several detectors will be built for detecting the particles produced by collisions in the LHC. The detectors have different parts that detect different types of particle. The hadron calorimeter, for example, can only detect hadrons.

Formulae

MANY OF THE PRINCIPLES EXPLAINED IN THIS BOOK can be expressed as formulae. The use of symbols to represent different values enables the physicist to make quick calculations, reducing even complicated physical phenomena to simple mathematical formulae.

WEIGHT

Weight is equal to mass multiplied by acceleration due to gravity

W = mg

W = weight
m = mass
g = acceleration due to gravity

TURNING FORCE

Turning force is equal to force multiplied by distance of applied force from pivot

T = Fd

T = turning force (moment)
F = applied force
d = distance

PRESSURE

Pressure is equal to force applied divided by area over which force acts

P = F/A

P = pressure
F = applied force
A = area over which force acts

FORCE AND MOTION

NEWTON'S SECOND LAW

Acceleration is equal to force divided by mass

a = F/m

SPEED

Speed is equal to distance divided by time

v = d/t

CONSTANT ACCELERATION

Acceleration is equal to change in speed divided by time taken for that change

$a = v_2 - v_1 / t$

MOMENTUM

Momentum is equal to mass multiplied by speed

p = mv

F = applied force
v_1 = speed of object one
v_2 = speed of object two
t = time
d = distance
p = momentum
m = mass
a = acceleration

GRAVITATION

Gravitational force equals a constant, multiplied by mass one, multiplied by mass two, divided by the distance between the masses squared

$F = Gm_1m_2/d^2$

F = gravitational force between two objects
G = gravitational constant
m_1 = mass of object one
m_2 = mass of object two
d = distance between the two objects

FRICTION

Frictional force between two surfaces is equal to the coefficient of friction multiplied by the force acting to keep the surfaces together

$F = \mu N$

F = frictional force
μ = coefficient of friction; this varies with materials
N = force between two surfaces

AIR RESISTANCE

Force is proportional to speed

$F \propto v$

F = force of resistance due to air
v = speed of motion through air

WORK

Work is equal to force multiplied by distance

W = Fd

W = work done
F = applied force
d = distance moved in line with force

CENTRIPETAL FORCE

Force is equal to mass multiplied by the speed squared divided by the radius

$F = mv^2/r$

F = centripetal force
m = mass of object
v = speed of circular motion
r = radius of object's path

LIQUID PRESSURE

Pressure is equal to the liquid's density multiplied by acceleration due to gravity multiplied by height of water above point

$P = \rho gh$

P = pressure
ρ = liquid density
g = acceleration due to gravity
h = height of water above measured point

ELASTICITY

The extension of a solid is proportional to the force applied to it

$F \propto x$

x = extension of solid
F = applied force

GAS LAWS

BOYLE'S LAW

Volume is proportional to one divided by pressure

$V \propto 1/P$

CHARLES' LAW

Volume is proportional to temperature

$V \propto T$

PRESSURE LAW

Pressure is proportional to temperature

$P \propto T$

THE IDEAL GAS EQUATION

Pressure multiplied by volume is equal to ideal gas constant multiplied by temperature

PV = RT

V = volume
P = pressure
T = temperature
R = the ideal gas constant

ELECTRIC CIRCUITS

OHM'S LAW

Current is equal to voltage divided by resistance

I = V/R

POWER

Power is equal to voltage multiplied by current

P = VI

I = current
V = voltage
R = resistance
P = power

IMAGE FORMATION

One divided by the focal length is equal to one divided by the object's distance from lens added to one divided by distance from the lens to the image

$1/f = 1/u + 1/v$

f = focal length
u = object's distance from lens
v = distance from lens to image

Appendix: useful data

PHYSICISTS USE STANDARD UNITS of measurement called SI units (Système International), which include the kilogram, the metre, and the second. In addition to these standard units, there are many other units of measurement. The tables below give details of these.

TEMPERATURE SCALES

To convert	Into	Equation
Celsius (C)	Fahrenheit (F)	F = (C x 9 ÷ 5) +32
Fahrenheit	Celsius	C = (F - 32) x 5 ÷ 9
Celsius	Kelvin (K)	K = C + 273
Kelvin	Celsius	C = K - 273
Fahrenheit	Kelvin	K = ((F - 32) x 5 ÷ 9) + 273

METRIC – IMPERIAL CONVERSIONS

To convert	Into	Multiply by
Length		
Centimetres	inches	0.3937
Metres	feet	3.281
Kilometres	miles	0.6214
Metres	yards	1.094
Mass		
Grams	ounces	0.03527
Kilograms	pounds	2.205
Tonnes	tons	1.102
Area		
Square centimetres	square inches	0.1550
Square metres	square feet	10.76
Hectares	acres	2.471
Square kilometres	square miles	0.3861
Square metres	square yards	1.196
Volume		
Cubic centimetres	cubic inches	0.06102
Cubic metres	cubic feet	35.31
Cubic metres	cubic yards	1.308
Capacity (liquid)		
Litres	pints	1.760
Litres	gallons	0.2200

IMPERIAL – METRIC CONVERSIONS

To convert	Into	Multiply by
Length		
Inches	centimetres	2.540
Feet	metres	0.3048
Miles	kilometres	1.609
Yards	metres	0.9144
Mass		
Ounces	grams	28.35
Pounds	kilograms	0.4536
Tons	tonnes	1.0160
Area		
Square inches	square centimetres	6.452
Square feet	square metres	0.09290
Acres	hectares	0.4047
Square miles	square kilometres	2.590
Square yards	square metres	0.8361
Volume		
Cubic inches	cubic centimetres	16.39
Cubic feet	cubic metres	0.02832
Cubic yards	cubic metres	0.7646
Capacity (liquid)		
Pints	litres	0.5683
Gallons	litres	4.546

PHYSICS SYMBOLS

Symbol	Meaning
α	alpha particle
β	beta particle
γ	gamma ray: photon
ϵ	electromotive force
η	efficiency; viscosity
λ	wavelength
μ	micro-; permeability
ν	frequency; neutrino
ρ	density; resistivity
σ	conductivity
c	speed of light

POWERS OF TEN

Factor	Name	Prefix	Symbol
10^{18}	quintillion	exa-	E
10^{15}	quadrillion	peta-	P
10^{12}	billion	tera-	T
10^{9}	thousand million	giga-	G
10^{6}	million	mega-	M
10^{3}	thousand	kilo-	k
10^{2}	hundred	hecto-	h
10^{1}	ten	deca-	da
10^{-1}	one tenth	deci-	d
10^{-2}	one hundredth	centi-	c
10^{-3}	one thousandth	milli-	m
10^{-6}	one millionth	micro-	μ
10^{-9}	one thousand millionth	nano-	n
10^{-12}	one billionth	pico-	p
10^{-15}	one quadrillionth	femto-	f
10^{-18}	one quintillionth	atto-	a

The above terms are those used in the UK. In the US, one thousand million is known as one billion and one million million is known as one trillion.

BASE SI UNITS

Physical quantity	SI unit	Symbol
Length	metre	m
Mass	kilogram	kg
Time	second	s
Electric current	ampere	A
Temperature	kelvin	K
Luminous intensity	candela	cd
Amount of substance	mole	mol
Plane angle	radian	rad
Solid angle	steradian	sr

DERIVED SI UNITS

Physical quantity	SI unit	Symbol
Frequency	hertz	Hz
Energy	joule	J
Force	newton	N
Power	watt	W
Pressure	pascal (newtons per square metre)	Pa (Nm^{-2})
Electric charge	coulomb	C
Voltage	volt	V
Electric resistance	ohm	Ω

Glossary

ABSOLUTE ZERO: The lowest possible **temperature**. The higher the temperature of matter, the more movement, or **kinetic energy**, its particles possess. At absolute zero the particles do not move at all. Absolute zero is zero **kelvin**, -273.15° **Celsius** or -459.67° **Fahrenheit**.

ACCELERATION: A change in the **speed** of an object. A reduction in speed is a negative acceleration, and is often called a deceleration. Acceleration is usually measured in ms^{-2} (metres per second per second, or metres per second squared).

ACHROMATIC DOUBLET: A system of two **lenses** that eliminates **chromatic aberration**. The two lenses are made of different types of glass.

ADDITIVE PROCESS: Combining light of different colours. When light of more than one colour enters the eye, the result is a colour that is different from each of the initial colours. This is due to human eyes having three types of **cone** cell. The brain combines the signals from each type of cone, and interprets the result.

ADHESIVE FORCES: The attractive **forces** between two different types of matter, such as water and glass. The balance between adhesive and **cohesive** forces determines whether the **meniscus** of a liquid will be upwards or downwards.

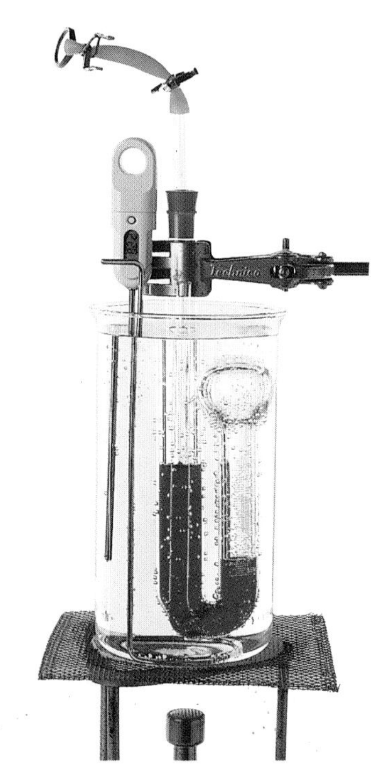

CHARLES' LAW

ALPHA DECAY/PARTICLE: The break-up of an unstable atomic **nucleus**, resulting in the release of a particle consisting of two **protons** and two **neutrons** – an alpha particle. During alpha decay, the **atomic number** of the nucleus reduces by two and the **atomic mass** by four (see **beta decay**).

AMORPHOUS SOLID: Any solid with particles not arranged in a regular, repeating pattern and therefore not composed of **crystals**. Because the particles are not regularly arranged, given sufficient time, an amorphous solid can flow, and is often called a supercooled liquid.

AMPLITUDE: The intensity of a wave motion. For a sound wave, the amplitude determines how loud the sound will be. For a water wave, the amplitude is the height of the wave, half the distance from peak to trough.

ANGULAR MOMENTUM The product of a spinning object's speed of rotation and its **moment** of **inertia**. An object's moment of inertia is a measure of how hard it is to set the object spinning.

LUBRICATION

ANODE: The positive electrode of any electrical apparatus. Because the anode is connected to the positive electrical supply, **electrons** are attracted to it. Anodes are used in X-ray tubes and **cathode ray tubes**.

ANTIPARTICLE: A particle that has the same **mass** as another particle, but has opposite charge or some other opposite property.

ATOM: A tiny particle. The building blocks of matter, atoms are the smallest part of an **element**. They are typically 10^{-10}m (one ten millionth of a millimetre) in diameter, and consist of a positively charged **nucleus** surrounded by negatively charged **electrons**.

ATOMIC MASS: The total mass of **protons** and **neutrons** in the **nucleus** of an **atom**, expressed in atomic mass units. Protons and neutrons each have a mass of one atomic mass unit. Fluorine-19, with nine protons and ten neutrons, has an atomic mass of nineteen.

ATOMIC NUMBER: The number of **protons** present in the **nucleus** of an **atom**.

BETA DECAY/PARTICLE: The break-up of an unstable atomic **nucleus**, resulting in the release of a fast-moving **electron**. This electron is called a beta particle. During beta decay, the **atomic number** of the nucleus actually increases by one, because a **neutron** changes into a **proton**, releasing an **electron**. The **atomic mass** is unchanged (see **alpha decay**).

BROWNIAN MOTION: The random motion of small solid objects, such as smoke particles, which can be observed under a microscope. The movement is caused by **atoms** and **molecules** of liquid or gas bombarding the solid objects.

BUBBLE CHAMBER: A device used to detect particles in collisions that take place in particle accelerators. The chamber contains a liquid, such as liquid hydrogen, held just below its boiling point. Any particles that have **electric charge** cause **atoms** in the liquid to become **ions.** The liquid boils around these ions, forming tiny gas bubbles wherever a charged particle passes. A strong **magnetic field** in the chamber causes the particles to travel in curved tracks, and the particle types can be identified by their tracks.

CAPILLARY ACTION: The rising or falling of a liquid in a narrow tube, above or below the liquid surface, due to surface tension. If **adhesive forces** are stronger than **cohesive forces**, as in the case of water in glass tubes, the liquid will climb up the tube. The narrower the tube, the higher the liquid will rise or fall.

CATHODE: The negative electrode of any electrical apparatus. Because the cathode is connected to the negative electrical supply, **electrons** are pushed away from it (see **anode**).

CATHODE RAY TUBE: A sealed glass tube used in the display of most televisions. Inside the tube, **electrons** leave a **cathode**, and are attracted towards the high-**voltage anode**. The electrons form a beam, sometimes called a cathode ray, which can be observed as it touches a **phosphorescent** screen.

GYROSCOPE

CELSIUS: A **temperature** scale on which water freezes at zero degrees and boils at 100 degrees. Each degree celsius is equal to one degree **kelvin**, but the kelvin scale begins at absolute zero (-273.15°C). Once called the Centigrade scale, the name was changed in 1948.

CENTRE OF GRAVITY: The point of an object at which clockwise and anti-clockwise **moments** are equal and the object therefore balances.

CENTRIPETAL FORCE: The **force** needed to keep an object moving in a circle or an **ellipse.** In the case of circular motion, the force is always directed to the centre of the circle, and depends upon the **mass** and **speed** of the object and the radius of the circle.

CERN (CONSEIL EUROPEEN POUR LA RECHERCHE NUCLEAIRE): The European Laboratory for Nuclear Physics, near Geneva on the Swiss-French border run by nineteen European nations.

CHAIN REACTION: A process, such as nuclear fission, in which each reaction is in turn the stimulus of a further reaction.

CHARGE: See **electric charge**.

CHROMATIC ABERRATION: A defect in a **lens**, caused by the fact that different **wavelengths** of light refract by different amounts as they pass through glass. The result of the defect is that different colours of light focus at different points. An image produced by the lens therefore has coloured fringes around it. The problem can be solved by using an **achromatic doublet**.

CLOUD CHAMBER: A device used to detect and track particles resulting from radioactive decay. It is a sealed unit containing a vapour, usually alcohol, just at the

FOCUSING AN IMAGE

point of condensing to form a liquid. **Alpha** and **beta** particles possess **electric charge**, and for this reason provide sites around which the vapour can condense. The tracks of the particles appear as paths of tiny droplets (see **bubble chamber**).

COHESIVE FORCES
The attractive **forces** between atoms or molecules in a liquid, such as water. Cohesive forces are responsible for **surface tension** (see **adhesive forces**).

COMPONENT
The effect of a **force** in a particular direction. A force can be thought of as a combination of two or more components.

COMPRESSION
The action of quashing a substance, so that it takes up a smaller space. When a **gas** is compressed, its **pressure** increases. When a solid is compressed, reaction forces are produced. These forces are responsible for the strength of a solid such as concrete, which is said to be "strong in compression".

CONCAVE: Shaped like the inside of a bowl. Concave mirrors make parallel light **converge**. Concave **lenses** make parallel light **diverge**.

CONDUCTIVE: Describes a material that allows **electric current** to flow through it easily. A material with a high conductivity allows electricity to flow easily, and is called a conductor. The term can be used to describe heat flow as well as the flow of electricity.

CONES: Cells at the back of the human eye (the retina), which are sensitive to light of particular ranges of colour. The cones allow for colour vision. There are three types of cone cell: red-, green-, and blue-sensitive.

CONSTRUCTIVE INTERFERENCE: The combination of two waves where the waves are "in step". Hence the peaks of one wave correspond to the peaks of the other. The **amplitude** of the combined wave is the sum of the amplitudes of the individual waves.

CONVERGE: To come together, as parallel light does, when it comes to a point of focus.

CONVEX: Shaped like the outside of a bowl. Convex **lenses** make parallel light **converge**. Convex mirrors make parallel light **diverge**.

CRITICAL ANGLE: The angle at or above which light striking the boundary between two different materials undergoes **total internal reflection**.

CRYOGENIC UNIT: Device used to reduce the temperature of substances to very low values, often only a few degrees above **absolute zero**. Cryogenic units are used in particle accelerators, such as those at **CERN**, to produce liquid helium, which cools **electromagnets** necessary for the operation of the accelerator.

CRYSTAL LATTICE: A regular, repeating arrangement of **atoms** or **molecules** in a solid (see **unit cell**).

CRYSTALS: Solids whose atoms or molecules are arranged in a **crystal lattice**.

CURRENT: See **electric current**.

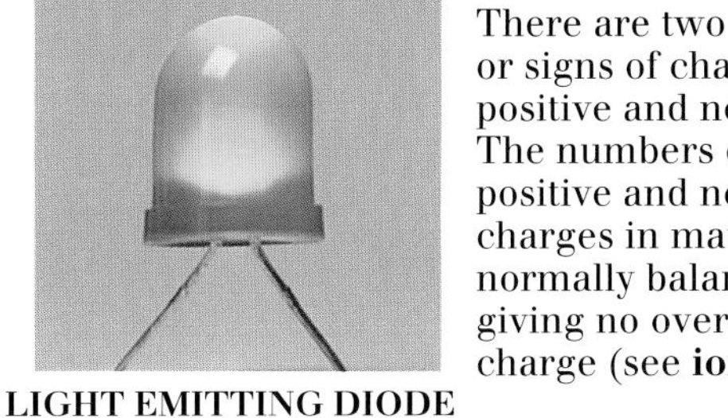

LIGHT EMITTING DIODE

DENSITY: A measure of the concentration of **mass** in a substance. The numerical value for density is calculated by dividing the mass of a given amount of the substance by its **volume**.

DESTRUCTIVE INTERFERENCE: Combination of two waves where the waves are "out of step". This means that the peaks of one wave correspond to the troughs of the other. The **amplitude** of the combined wave is therefore reduced.

DIFFRACTION: The bending of waves around the edge of an object. A small gap consists of two edges, so the waves are bent into concentric curves.

DIFFUSION: The mixing of substances, caused by the random motion of their particles. Diffusion is most noticeable in **gases**, because the movement of the particles is much faster than in solids or liquids.

DISPLACEMENT: A movement away or the distance of an object from its normal position.

DIVERGE: To move apart, as parallel light does when it passes through a concave **lens**.

DOMAINS: Tiny regions, between 0.1 and 1 mm across, within magnetic materials. Every **atom** of a magnetic substance, such as iron, is itself a tiny magnet. All of the atoms within a given domain are lined up, so that each domain is magnetized in a particular direction. In an unmagnetized state, the domains cancel each other out. When a material is magnetized, the domains are made to line up with each other.

ELASTICITY: The ability of a substance to regain its size and shape after being stretched by **forces** of **tension**. Forces of attraction between **atoms** within the substance are made stronger when the atoms are pulled apart. These forces are responsible for elasticity.

ELECTRIC CHARGE: A property of certain particles or substances that results in **electrostatic forces**. There are two types or signs of charge – positive and negative. The numbers of positive and negative charges in matter is normally balanced, giving no overall charge (see **ion**).

ELECTRIC CURRENT: The movement of particles with **electric charge**. Most electric currents are the result of moving **electrons**. The movement of electrons is caused by **electrostatic** or **electromagnetic forces**.

ELECTRIC FIELD: A region in which a particle with **electric charge** will experience an **electrostatic force**.

ELECTRODE: Part of an electrical device connected to the electrical supply. The positive electrode is called the **anode**, while the negative electrode is called the **cathode**.

ELECTROMAGNET: A device made by winding a continuous coil of wire around an iron core. **Electric current** flowing through the wire creates magnetism that lines up the **domains** in the iron, turning it into a temporary magnet.

ELECTROMAGNETIC FORCES: The **forces** on **electric charges** moving in a **magnetic field**. The size and direction of the force depends upon the speed, sign, and size of the charge, and on the strength and direction of the magnetic field.

ELECTROMAGNETIC RADIATION: A form of energy that travels through space and matter. It is associated with **electric fields** and **magnetic fields**, and can be thought of as a wave motion involving these fields. It can also be thought of as a stream of particles called **photons**. The many types of radiation, include light, radio waves, and X-rays.

ELECTROMAGNETIC SPECTRUM: The range of **electromagnetic radiation**. Each type of radiation is identical except for its **wavelength** and its energy. Radiation types with short wavelengths and high energy include X-rays and gamma rays, while longer-wavelength, lower-energy radiation includes infrared and radio waves.

ELECTRON: A particle found in all **atoms**. All electrons have one unit of negative **electric charge**.

ELECTROSCOPE: An instrument for measuring the extent of imbalanced **electric charge** in an object. The most common example is a glass box with two pieces of gold foil that are pushed apart as they are charged by **induction**.

ELECTROSTATIC FORCES: The forces between electric charges. Two charges of the same sign will push apart, or repel. Charges of different sign pull together, or attract.

ELEMENT: A substance consisting of only one type of **atom**. Examples are hydrogen, oxygen, and fluorine.

ELLIPSE: A shape that looks like a flattened circle. The orbits of the planets are ellipses.

EQUILIBRIUM: A balanced state at which the **resultant** of a number of **forces** on an object is zero.

ERROR BAR: A vertical or horizontal line drawn on a **graph** to indicate the margin of accuracy with which a particular measurement was taken.

EVAPORATION: The loss of **atoms** or **molecules** from a liquid, as they break free of the liquid to become a vapour. Evaporation takes place below the boiling temperature of the liquid.

SOAP BUBBLE

INDUCTION

EXCITED: In possession of extra energy. **Electrons** in **atoms** can be excited by heat or light energy. When they do so, they occupy a new position in the atom, according to their new energy.

FAHRENHEIT: Scale of **temperature** on which water freezes at 32 degrees and boils at 212 degrees.

FILAMENT: The fine wire in an incandescent light bulb. The filament heats up when electric current flows through it. At high temperatures, the filament glows.

FISSION: The splitting of unstable **nuclei** of **atoms**. The process begins as a free **neutron** joins the nucleus, making it more unstable. The nucleus splits into two smaller, more stable nuclei and releases further free neutrons and a large amount of energy in a **chain reaction**.

FLUID: Any substance that flows. Liquids and gases are both fluids.

FLUORESCENCE: A type of **luminescence** in which a substance glows with visible light immediately after being **excited** by invisible ultraviolet radiation.

FOCAL LENGTH: The distance from a **lens** or curved mirror at which a parallel beam of light comes to a focus.

FORCE: A push or a pull.

FREQUENCY: The regularity with which something happens. It is most often applied to a wave or vibration. The number of times the complete cycle of a wave happens each second is the frequency of the wave, and is measured in hertz (Hz).

FULCRUM: The point about which an object turns. For example, the fulcrum of a lever is its pivot.

FUSION: A joining of small **nuclei** of **atoms** to form larger nuclei. In some cases, such as when hydrogen atoms fuse to form atoms of the **element** helium, there is a huge release of energy during the process.

GAUGE BOSON: A particle exchanged between two interacting particles. At the sub-microscopic level of the tiniest particles, the exchange is responsible for the four forces: **gravity**, **electromagnetic force**, the **weak interaction**, and the **strong nuclear force**.

GEIGER-MULLER TUBE: A device for detecting radioactivity. An **electric current** flows between the wall of the tube and a metal wire at its centre when **alpha** or **beta particles** enter.

GENERATOR: A machine that produces an electrical **voltage** whenever its rotor is turned. The **kinetic energy** of the rotor becomes electrical energy because of the presence of coils and magnets.

GLUONS: According to modern theory particles responsible for carrying the **strong nuclear force** (see **gauge boson**).

GRAPH: A visual representation of a set of results of an experiment. A graph will highlight any relationships between the various types of data.

GRAVITY/GRAVITATION: A force of attraction between all objects with mass. The size of the force depends upon the masses of the two objects and the distance between the objects. Modern theory says that gravity is carried by particles known as gravitons (see **gauge boson**).

MENISCUS

GYROSCOPE
Usually thought of as a spinning metal disk supported in a metal cage, the word can refer to any spinning object. Gyroscopes have stability because they spin.

HADRON
Any particle that is composed of **quarks**. Examples are the **proton** and the pi **meson**.

HIGGS BOSON: Hypothetical particle whose existence would link the **electromagnetic force** and the **weak interaction**, and explain why particles have **mass**.

HOLE: A vacant electron position within the **crystal lattice** of a **semiconductor** that can be thought of as a positive charge.

IMAGE: A picture formed by a lens or a curved mirror. Images cast on screens by convex lenses are called real images, while those seen using telescopes or microscopes, which cannot be directly projected, are called virtual images.

INDUCTION: 1. The apparent charging of one object by an electrically charged object nearby. The charging is apparent, since it is only a shift of **electric charge** within the object. 2. The magnetization of iron objects in the presence of a magnet. The **domains** inside the iron line up with the **magnetic field** of the magnet.

INERTIA: The resistance of an object to any change in its motion.

INFRARED RADIATION: A type of **electromagnetic radiation**, with a **wavelength** shorter than visible light.

INTERFERENCE: The combination of two or more waves.

ION: An **atom** with an overall electric charge. The numbers of positively charged protons and of negatively charged electrons in an atom are normally equal. But with the removal of one or more electrons the atom is left with a net positive charge, while extra electrons give a net negative charge.

KELVIN: The absolute scale of **temperature**, the Kelvin scale begins at absolute zero, and unlike the **Celsius** and **Fahrenheit** scales, does not rely on fixed points.

KINETIC ENERGY: The energy of a moving object, dependent upon the mass and speed of the object.

LATENT HEAT: Heat energy that melts a solid or vaporizes a liquid. Latent heat does not raise the temperature of the substance.

LENS: A curved piece of glass or other transparent material that refracts light, and can form **images**.

LIMITING FRICTION: The **force** which must be overcome to start an object moving when it is in contact with a surface.

LUMINESCENCE: The emission of light due to a decrease in the energy level of an **excited electron** within an **atom** or **molecule**. The two main types are **fluorescence** and **phosphorescence**.

MAGNETIC FIELD: A field of **force** around a magnet's poles or around a wire carrying an **electric current**.

MASS: The measure of an object's **inertia**. Mass is also defined in terms of **gravitation**. The gravitational force between two objects depends upon their masses.

MELTING POINT: The temperature at which a solid substance becomes a liquid. It is dependent upon atmospheric **pressure**.

MENISCUS: The curved surface of a liquid where it meets its container. It is caused by a combination of **adhesive** and **cohesive forces**.

MESONS: A **hadron** consisting of two quarks. An example is the pi meson, which carries the **strong nuclear force** between **protons** and **neutrons** within the **nucleus**.

MICROMETER: A device used to measure very small displacements.

MOLECULE: The smallest amount of a compound. A water molecule consists of two **atoms** of hydrogen and one of oxygen.

MOMENT: The turning effect of a **force**.

NEUTRON: One of the particles in the **nucleus** of an **atom**. It is a **hadron**, and has no **electric charge**.

NEWTON METER: A device used to measure **force**. A pointer moves along a scale as a spring inside the meter extends. The extension of the spring depends upon the applied force.

NUCLEAR REACTIONS: Changes involving the **nuclei** of **atoms**, such as **fission** and **fusion**.

NUCLEUS: The central part of an **atom**. It has a positive **electric charge** because it contains **protons**.

ORBITALS: Regions of an **atom** in which **electrons** are found. The name comes from the word "orbit", since electrons were originally thought to follow definite paths around the **nucleus**.

OSCILLATOR: An electric circuit that produces an alternating **electric current**, which repeatedly changes direction.

PERMANENT MAGNETS: Objects with a fixed magnetism. The **domains** in a permanent magnet always align to produce a **magnetic field** (see **electromagnet**).

MAGNETIC FIELD

PHOSPHORESCENCE (PHOSPHOR): A type of **luminescence** in which a substance glows with visible light some time after being **excited** (see **fluorescence**). A phosphor is any substance exhibiting phosphorescence.

PHOTON: A particle of **electromagnetic radiation.** The **energy** of a photon depends only upon the **wavelength** of the radiation. A photon can be thought of as a packet of waves.

PLANE WAVE: A wave motion in which the waves are parallel to one another and perpendicular to the direction of the wave's motion.

POSITRON: The **antiparticle** of the **electron.** It is identical to the electron in every way, except that it has positive **electric charge.**

POTENTIAL ENERGY: Energy that is "stored" in some way. For example, an object held in the air has potential energy by virtue of its height and the **gravitational force** pulling it downwards.

PRESSURE: A measure of the concentration of a **force.** The pressure exerted by a force is equal to the size of the force divided by the area over which it acts. Solids, liquids and gases exert pressure.

PRIMARY COLOURS: A set of three colours, which, when combined in the correct proportion, can produce any other colour. The set of primaries for the **additive process** is different from that for the **subtractive process.**

PRINCIPLE OF SUPERPOSITION: The rules governing the **interference** of waves.

PRINCIPLE OF THE CONSERVATION OF ENERGY: Energy can be neither created nor destroyed, it can only change, or transfer, from one form to another.

1 KG MASS

PROTON: A **hadron** found within the **nucleus** of an **atom.** It has a positive **electric charge.**

QUARKS: Particles that combine together to form **hadrons,** such as **protons** and **neutrons.** No quark has ever been detected in isolation.

RAREFACTION: The lowering of the **density** and **pressure** of a gas; the opposite of **compression.**

REACTION: A **force** produced by an object that is equal and opposite to a force applied to the object.

REFRACTION: The bending of light or other **electromagnetic radiation** as it passes from one material to another.

RESISTANCE: A measure of the opposition to the flow of **electric current.** It is the ratio of **voltage** to current.

RESULTANT: The combined effect of two or more **forces.**

SANKEY DIAGRAM: An illustration of the energy changes in a process. The diagram consists of a large arrow, which represents the input of energy to the process, and which splits according to the energy changes that occur.

SEMICONDUCTOR: A material in which the **electrons** are held only loosely to their **atoms.** The electrons can become free, and the material therefore becomes **conductive,** with only a small input of **energy.**

AN INCLINED PLANE

SHELL: An energy level occupied by **electrons** within an **atom.** It is generally true that the lower the energy of electrons in the shell, the closer it is to the **nucleus.**

SI UNITS (SYSTEME INTERNATIONAL D'UNITES): A system of units accepted by the world-wide scientific community as the standard. Its seven base units include the kilogram and the second.

SOLENOID: A long coil of wire that produces a **magnetic field** similar to that of a bar magnet. With an iron bar inside the coil, a solenoid becomes an **electromagnet.**

SOLUTION: A mixture of two substances, in which the particles of the substances are uniformly mixed, for example salt dissolved in water.

SPEED: The rate at which an object moves, equal to the distance moved divided by the time taken.

STATE: The form of a substance, either solid, liquid, or gas.

STRONG NUCLEAR FORCE: The **force** between **hadrons.** It is carried by **gluons,** or by combinations of **quarks** (see **gauge boson**). The strong nuclear force is responsible for holding the **nucleus** together.

SUBTRACTIVE PROCESS: The process by which pigments absorb parts of the visible spectrum of light, but reflect others, making objects appear to have colour.

SUPERCOOLED LIQUID: See **amorphous solid.**

SURFACE TENSION: The **resultant force** at the surface of a liquid, due to the **cohesive forces** between the particles of the liquid.

TEMPERATURE: How hot or cold a substance is. Temperature relates directly to the **kinetic energy** of a substance's particles: the particles in hotter objects have more **kinetic energy.**

TENSION: A **reaction force** in a solid that is stretched, which pulls the **atoms** of the solid together. It is the opposite of **compression.**

TERMINAL VELOCITY: The maximum speed attained by an object falling through a liquid or gas. A parachute falling through air has a relatively low terminal velocity, while that of a ball bearing will be much greater.

THERMAL EXPANSION: The expansion of a solid as its **temperature** increases. It is due to the increased vibration of the **atoms** and **molecules** of the solid. This increased vibration occurs at higher temperatures, due to the increased **kinetic energy** of the atoms and molecules.

TOTAL INTERNAL REFLECTION: The reflection of light from the border between two materials, as the light passes from the denser to the less dense material.

TURBINE: A machine in which a liquid or a gas causes rotation. When attached to a **generator,** the turning of the turbine helps to generate electricity.

UNIT CELL: The group of **atoms** or **molecules** in a crystal; when repeated, it forms the **crystal lattice.** There are seven naturally occuring unit cell types.

MEASURING RESISTANCE

UPTHRUST: An upward **force** on an object immersed in a liquid or a gas. Upthrust is the **resultant** of the liquid or gas pressure acting on the object. Upthrust supports ships in the ocean and hot-air balloons in the air.

VELOCITY: The **speed** and direction of an object's motion.

VERNIER SCALE: A scale attached to instruments such as calipers, which allows very accurate measurements to be taken.

VOLTAGE: A measure of the **force** on particles with **electric charge.** The voltage in an **electric circuit** pushes **electrons** around the circuit.

VOLUME: The amount of space an object takes up. This is measured in cubic metres (m^3).

WAVELENGTH: The distance from one wave peak to another. The wavelength of **electromagnetic radiation** determines the type of radiation. For example, X-rays have a shorter wavelength than light. Different wavelengths of light cause the sensation of colour.

THE WEAK INTERACTION: A **force** between some types of particle, including **electrons.** Also involved in the decay of **hadrons,** such as the **beta decay** of **neutrons** in the **nucleus.** The force is carried by W and Z particles (see **gauge boson**).

WEIGHT: The **force** of **gravity** on an object. It is dependent on the **mass** of the object. Weight is therefore variable under different gravitational conditions.

WORK: The amount of energy involved in a particular task. For example, work is said to be done when a pulley lifts a load. The amount of work done is equal to the force acting multiplied by the distance moved.

Index

F

G

H

I

J

K

L

M

N

O

P

Q

R

S

T

U

V

Acknowledgements

Dorling Kindersley would like to thank:
Griffin and George, Loughborough, for the loan of scientific equipment; University College London, for the loan of glassware; Maplin electronics, Hammersmith

Special thanks to Lew Instone at Griffin and George, Peter Leighton at UCL, and Patrick Rolleston at Kensington Park School

Additional design assistance:
Carla De Abreu, Anthea Forle

Additional editorial assistance:
Louise Candlish, Phillippa Colvin, Caroline Hunt, Jane Mason, Jane Sarluis, Roger Tritton

Additional photography:
Tim Ridley

Model making:
Simon Murrell

Picture research:
Anna Lord

Picture credits:
CERN 50bl, 52br; Simon Fraser 37tr; Robert Harding/C. Delli 37cr; Image Bank/ Mel Digiacomo 47crb; Bruce Iverson 44c; Newage International Ltd 36tc; Omikron 52cr; David Parker 39tc; Des Reid/Marian Tully 24tr; John Sanford 45tr; Science Photo Library/Lawrence Berkeley Laboratory 52tr; Stammers/Thompson 39bc; Zefa 29bc, 35cl, 37c
(t=top, b=bottom, c=centre, l=left, r=right)